Student Resource Manual

Maricopa Mathematics Modules

Student Resource Manual

Maricopa Mathematics Modules

Alan Jacobs
Keith Worth
Scott Adamson
Paula Cheslik
Anne Dudley
David Dudley
Teri Glaess
Karen Hay

Houghton Mifflin Company Boston New York

Senior Sponsoring Editor: Maureen O'Connor
Development Editor: Dawn Nuttall
Senior Manufacturing Coordinator: Sally Culler
Marketing Manager: Ros Kane
Editorial Associate: Amanda Bafaro

Printed in the U.S.A.

ISBN: 0-618-00050-X
123456789-B+B-04 03 02 01 00

Student Resource Manual
Table of Contents

To the Student

Mathematics instruction at every level should propel students into future success as citizens, employees, employers, and as lifelong learners.

Welcome to the Maricopa Mathematics Modules! We value and respect you as a student and hope you have a rich and rewarding experience. These modules are the result of "rethinking" how the learning of mathematics should best happen. As a result, our modules no longer focus on the teacher's activity; rather they focus on what you, the learner, does to learn. You will learn by doing activities in a meaningful context using suitable technologies. Contexts that have meaning arise naturally from client disciplines such as science and business as well as quantitative presentations and arguments in newspapers and other media.

The philosophy behind these modules is to value the goal of the successful learner, you. Ideally, the larger and more important goal, for you as a successful learner and user of mathematics, should be ownership of the relevant mathematical ideas and mathematical power. Your goal is not just passing a math class or getting a requirement completed for graduation. Instead, your larger goal should be to ***"C-R-E-A-T-E"*** your own mathematics. We have coined the ***"C-R-E-A-T-E"*** acronym to describe the five learner outcomes below:

Connect **:**
Learners employ a variety of methods to represent and explore mathematical ideas. They construct and apply models that connect mathematics to the world.

Reason **:**
Learners demonstrate clear reasoning in analyzing information.

Express **:**
Learners read, write, listen to, and speak mathematics both individually and in teams.

Appreciate **:**
Learners value the power of mathematics. They are confident, flexible and persistent lifelong learners and users of mathematics.

Tap into Technology **:**
Learners discover the benefits and limitations of current technologies. They utilize technologies as resources for learning and problem solving.

Establish a Foundation :
Learners acquire and develop a core of content-specific knowledge and abilities, as well as strategies for learning.

We believe that all students should ***"C-R-E-A-T-E"*** mathematics!

"I have never taken a math course that actually requires you to think about the answers! Most classes focus on memorizing equations and steps; it always seemed there was very little logic in what you were doing. It made math highly anxiety provoking. This class was a learning experience for me. I feel my anxiety level has gone down. This class gave me confidence." – Student who learned using The Maricopa Mathematics Modules

A different approach to mathematics instruction means a different approach by you the student. As you study mathematics using the Maricopa Mathematics Modules, you will become increasingly aware of their benefits to your learning. You will be asked to discover, invent, and develop mathematical ideas. You will be asked to interpret and relate these ideas to physical contexts in the world around you.

Many of the modules' context-based problems lend themselves to the unique problem-solving power of a team approach, as do the important problems to solve in the workplace of today's modern world. Teaming skills will be something that you will refine through your experiences with our modules. Frequently, teaming with other students will be necessary to tackle larger, multi-faceted problems; each team member brings his/her own unique combination of learning and communicating styles to the problem-solving process. There is an obvious benefit for you from this experience when you take these skills to the modern workplace.

Overall, the modules will ask you to employ what will most probably be a different set of learning behaviors and study skills that you have had to apply to past learning experiences in mathematics. In many ways these learning behaviors and study skills are more closely related to the holistic approach required in other college disciplines for which a substantial element of critical thinking and problem solving is necessary.

"I liked the fact that instead of trying to learn tons of math concepts in a short amount of time, we focused on learning a few important concepts in depth, so we have a better understanding of those concepts so we can use that foundation to understand the other concepts. It is more efficient than trying to learn a bunch of miscellaneous concepts you will just forget in a few days." – Another student who learned using The Maricopa Mathematics Modules

"I liked the way it was taught, at first I didn't like figuring out my own formulas but then it really helped me in my math." – Another student who learned using The Maricopa Mathematics Modules

Now let's look at the learning behaviors and study skills successful students usually employ when learning mathematics using our modules.

Successful students…understand that they are in charge of their learning. As a result, successful students…

- make attendance in every class a high priority,
- find partners to study with,
- read, write, and then do…that is, they read about the mathematical ideas and contexts, answer all relevant thought questions, then do the activities and problems,
- share their team's goal and work as a contributing team member,
- communicate frequently with the teacher about mathematical ideas in the class,
- realize the importance of the holistic nature of a real, lasting learning experience in mathematics, so:
  - … they know that mechanical skills and algorithms are certainly necessary but in no way sufficient,
  - … they also know that understanding both the mathematical ideas and the relationships to relevant contexts are required for a lasting learning experience.
- take total ownership of their learning experience and their performance in the class.

We believe the Maricopa Mathematics Modules will probably be a different kind of experience than you have had in previous mathematics classrooms. We invite you to bring your outside experience and knowledge to our mathematics classroom. Our modules value and respect what you bring to the learning experience and provide you the opportunity to leave the experience with an ownership and power that comes from creating and constructing your own set of mathematical tools through your own learning experience! We truly believe that your commitment to doing your best work and employing your best learning behaviors and study skills will be rewarded with a lifelong change in the way you view and use mathematics!

Scott Adamson
Chandler Gilbert Community College

Paula Cheslik
Fountain Hills High School

Anne Dudley
Glendale Community College

David Dudley
Phoenix College

Teri Glaess
Scottsdale Community College

Karen Hay
Carnegie Mellon University

Alan Jacobs
Scottsdale Community College

Keith Worth
Scottsdale Community College

Beat Ratios and Juggling Proportions

Lesson 1 Juggling and Mathematics

Homework

1. 8 am 12 noon 4 pm 8 pm 12 mid 4 am 8 am

a. Time on-line : total shift time = 12.5 : 14.5

$$= \frac{12.5}{14.5} = \frac{25}{29}$$

≈ 0.862, by division

or 86.2%

The machine is on-line for approximately 86.2% of the shift time.

b. Time on-line: total time = 12.5 : 24

$$= \frac{12.5}{24} = \frac{25}{48}$$

≈ 0.521, by division

or 52.1%

The machine is on-line for approximately 52.1% of the day.

Lesson 3 Period and Frequency

Nitty Gritty: Solving Proportions

1. $\frac{12}{9} = \frac{x}{3}$

By inspection

$x = 4$

3. $\frac{7}{x} = \frac{21}{9}$

By inspection

$x = 3$

5. $\frac{2x}{5} = \frac{16}{9}$
multiply by 45
$45 \cdot \frac{2x}{5} = 45 \cdot \frac{16}{9}$
$9 \cdot (2x) = 5 \cdot 16$
$18x = 80$
divide by 18
$x = \frac{80}{18} = \frac{40}{9} = 4.4\overline{4}$

7. $\frac{x+2}{5} = \frac{16}{10}$
multiply by 10
$10 \cdot \frac{x+2}{5} = 10 \cdot \frac{16}{10}$
$2(x+2) = 16$
$2x + 4 = 16$
$-4 \quad -4$
$2x = 12$
divide by 2
$x = 6$

9. $\frac{5}{12} = \frac{x+7.5}{42}$
multiply by 84
$84 \cdot \frac{5}{12} = 84 \cdot \frac{x+7.5}{42}$
$7 \cdot 5 = 2(x+7.5)$
$2x + 15 = 35$
-15
$2x = 20$
divide by 2
$x = 10$

11. $\frac{17}{36} = \frac{3}{2x+5}$
multiply by $36(2x + 5)$
$36(2x+5)\frac{17}{36} = 36(2x+5)\frac{3}{2x+5}$
$17(2x+5) = 36 \cdot 3$
$34x + 85 = 108$
$-85 \quad -85$
$34x = 23$
divide by 34
$x = \frac{23}{34} \approx 0.68$

13. $\frac{2.5 \text{ cups}}{48 \text{ cookies}} = \frac{x \text{ cups}}{84 \text{ cookies}}$
multiply by 336
$336 \cdot \frac{2.5}{48} = 336 \cdot \frac{x}{84}$
$7 \cdot 2.5 = 4x$
$4x = 17.5$
divide by 4
$x = 4.375$ cups

15. $\frac{1 \text{ gallon}}{440 \text{ ft}^2} = \frac{x \text{ gallon}}{1750 \text{ ft}^2}$
multiply by 77000
$77000 \cdot \frac{1}{440} = 77000 \cdot \frac{x}{1750}$
$175 = 44x$
divide by 44
$x = \frac{175}{44} \approx 3.98$ gallons
$3\frac{1}{2}$ gallons is not enough paint

Homework

1. The period of the wafer machine is four hours. The 4-hour period consists of 3.5 hours on-line and 0.5 hour off-line. The frequency is 6 times per day. The frequency can be computed by dividing the number of hours in a day (24) by the period (4), $24 \div 4 = 6$. The machine is taken off-line 6 times per day.

3. 1 week 2 weeks 3 weeks 4 weeks

Diagonal lines indicate a waxing moon (approaching or increasing).
The dotted pattern indicates a waning moon (going away or shrinking).
The full moon cycle repeats approximately every 28 days, so the period is approximately 28 days.
Since there are 365 days per year, an event with a 28-day period has a frequency of $365 \div 28 \approx 13$ times per year.

5. Every 3000 miles

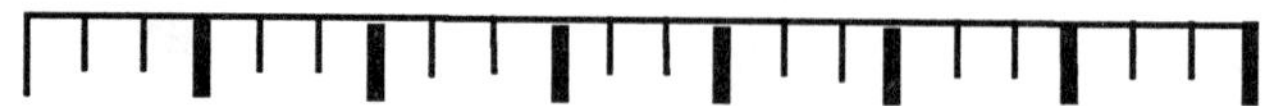

3000 miles 6000 miles 9000 miles 12,000 miles 15,000 miles 18,000 miles 21,000 miles

Solid indicates oil change
period: 3000 miles (note that the time may vary)
frequency: once every 3000 miles
Notice that the period and frequency are not expressed using units of time.

Every 3 months

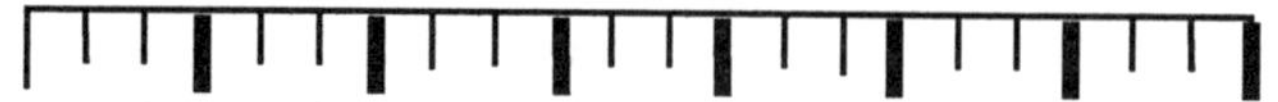

3 mo 6 mo 9 mo 12 mo 15 mo 18 mo 21 mo

Solid indicates oil change
period: 3 months (note that the miles may vary)
frequency: 4 times per year

7. elapsed time = (# of beats) · (period)
= 124 beats · 4 seconds per beat
= 496 seconds
= 8 minutes, 16 seconds

8. elapsed time = (# of beats) · (period)
= 372 beats · 4 seconds per beat
= 1488 seconds
= 24 minutes, 48 seconds

11. # of vibrations per hour = (# of vibrations per second) · (seconds per hour)
= 9,192,631,770 · (60 seconds per minute) · (60 minutes per hour)
= 33,093,474,370,000 or 3.3×10^{13} vibrations per hour.

Lesson 4 Phase Shifts

Homework

4. a. The period for machine B (gizmo maker) is 10 minutes. Machine A (widget builder) needs a gismo every two minutes. In other words, machine A needs 5 gizmos in 10 minutes. Therefore, it would take five B machines supplying gizmos to one A machine to ensure that A machine is not waiting for parts. The B machines should be shifted by two minutes.

Lesson 5 In Sync

Homework

1. a. 6 seconds & 14 seconds
   $6 \cdot 14 = 84$
   Common factors: 2
   LCM: $84/2 = 42$

   d. 20 seconds & 15 seconds
   $20 \cdot 15 = 300$
   Common factors: 5
   LCM: $300/5 = 60$

2. a. 3 seconds, 5 seconds & 6 seconds
   LCM of 3 and 5 is 15
   LCM of 15 and 5 is 30
   30 seconds

5. a. Recall that timelines are in sync at the l.c.m. of the timeline periods. The l.c.m. of 3 (machine A period) and 7 (machine B period) is $3 \cdot 7 = 21$ seconds. Machines A and B will be in sync every 21 seconds.

6. a. The l.c.m. of 6 (machine A period) and 8 (machine B period) is $2^3 \cdot 3 = 24$. The machines will produce chips in sync every 24 seconds.

7. a. The l.c.m. of 3 (machine A period) and 9 (machine B period) is $3^2 = 9$. The machines will produce chips in sync every 9 seconds.

Lesson 6

Homework

1. For each of the following pairs of beat frequencies, determine their common frequency and common period.

   a. $10 \dfrac{\text{beats}}{\text{min}}$ and $5 \dfrac{\text{beats}}{\text{min}}$
   common frequency = GCF = 5 beats/minute
   common period = (unit of time) ÷ (common frequency)
   = (60 seconds per minute) ÷ (5 beats per minute)
   = 12 seconds per beat

   c. $12 \dfrac{\text{beats}}{\text{min}}$ and $18 \dfrac{\text{beats}}{\text{min}}$
   common frequency = GCF = 6 beats/minute
   common period = (unit of time) ÷ (common frequency)
   = (60 seconds per minute) ÷ (6 beats per minute)
   = 10 seconds per beat

e. $5\ \frac{\text{beats}}{\text{min}}$ and $6\ \frac{\text{beats}}{\text{min}}$

Since 5 and 6 have no GCF, it will be better to find the common period and use it to calculate the common frequency.

First, the period for each frequency is found.

(period #1) = (unit of time) ÷ (frequency #1)
= 60 seconds per minute ÷ 5 beats per minute
= 12 seconds per beat.

(period #2) = (unit of time) ÷ (frequency #2)
= 60 seconds per minute ÷ 6 beats per minute
= 10 seconds per beat

Now the periods can be used to find the common period

(common period) = LCM of 12 and 10, or 60 seconds per beat.

The common period can be used to find the common frequency,

(common frequency) = (unit of time) ÷ (common period)
= 60 seconds per minute ÷ 60 seconds per beat
= 1 beat per minute

2. a. common frequency is the GCF of 4, 6, and 12: 2 beats/minute
common period = (unit of time) ÷ (common frequency)
= (60 seconds per minute) ÷ (2 beats per minute)
= 30 seconds per beat

c. common frequency is the GCF of 12, 16, and 24: 4 beats/minute
common period = (unit of time) ÷ (common frequency)
= (60 seconds per minute) ÷ (4 beats per minute)
= 15 seconds per beat

4. a. common frequency is the GCF of 4 and 6: 2 chips per minute
common period = (unit of time) ÷ (common frequency)
= 60 seconds per minute ÷ 2 chips per minute
= 30 seconds per chip

5. a. Common frequency is the GCF of 20 and 30: 10 chips per minute
common period = (unit of time) ÷ (common frequency)
= 60 seconds per minute ÷ 10 chips per minute
= 6 seconds per chip

Another way to solve this problem is to find the periods first.

(Period #1) = (unit of time) ÷ (frequency #1)
= 60 seconds per minute ÷ 20 chips per minute
= 3 seconds per chip

(Period #2) = (unit of time) ÷ (frequency #2)
= 60 seconds per minute ÷ 30 chips per minute
= 2 seconds per chip

The common period is the LCM of 3 and 2: 6 seconds per chip

(common frequency) = (unit of time) ÷ (common period)
= 60 seconds per minute ÷ 6 seconds per chip
= 10 chips per minute

Lesson 7 Applications

Homework

1.

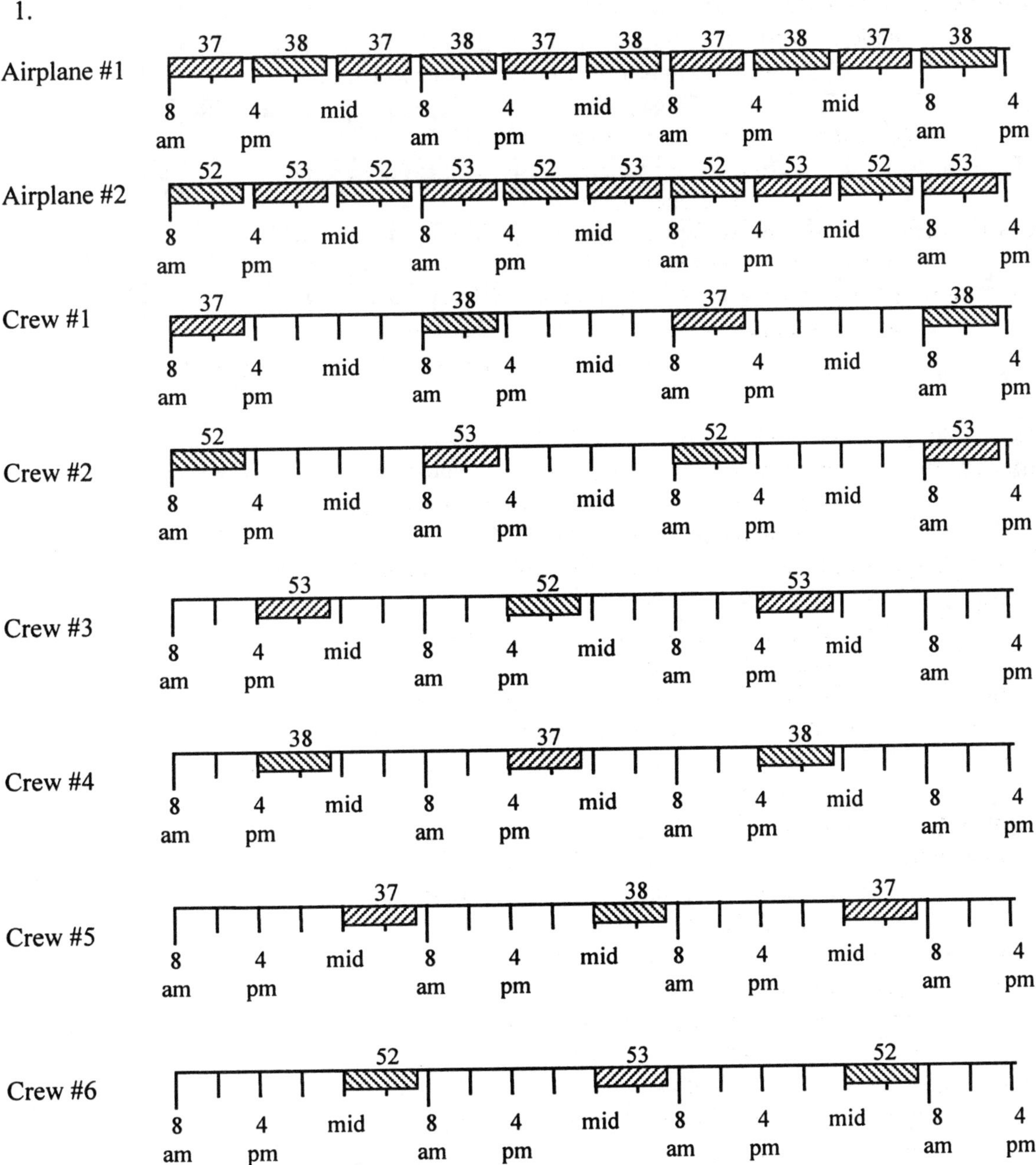

[NY to London bar] indicates flying from NY to London
[London to NY bar] indicates flying from London to NY

a. It will take six crews to fully staff this schedule.
b. Each crew gets 17 hours off.

2.

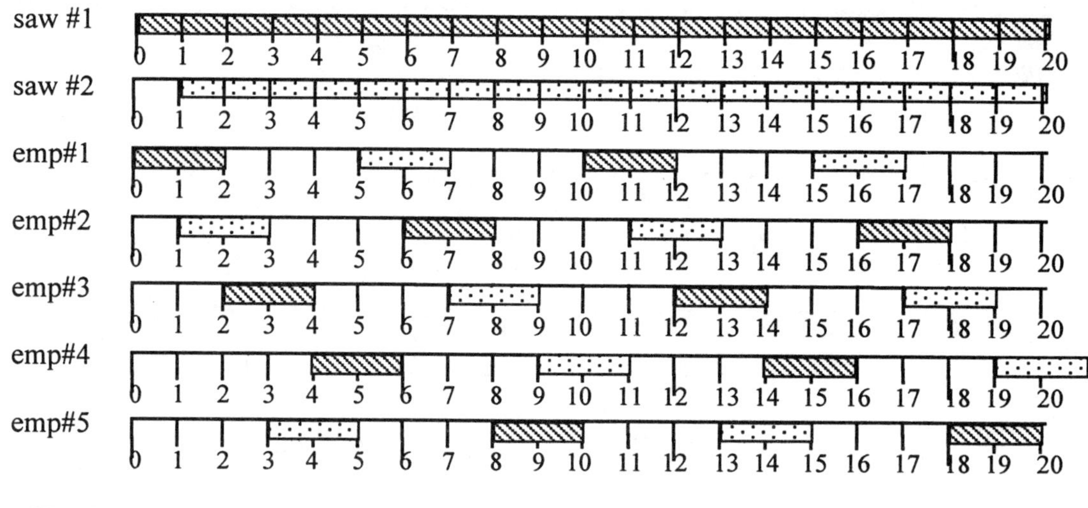

indicates saw #1

indicates saw #2

b. From the timeline it is clear that 5 employees are needed to staff the saws. The timelines were developed one by one. It may be easiest to first develop employee timelines to staff one of the saws. Then include shifts on the second saw, only adding more employees as needed.

d. Some things that may have been ignored while solving this problem include:
- time required for an employee to switch tasks
- coffee and lunch breaks
- bathroom breaks
- difficulties with the saw, mechanical failure or accident
- fire drills, power outages, etc.
- saw maintenance

Lesson 8 Juggling Schedules

Nitty Gritty: The Order of Operations

1. $6-3+5=$
   $3+5$
   8

3. $120\div4\bullet3=$
   $30\bullet3$
   90

5. $-4+6\bullet2^3=$
   $-4+6\bullet8$
   $-4+48$
   44

11. $21+5(3-10)^2=$
    $21+5(-7)^2$
    $21+5(49)$
    $21+245$
    266

13. $[91-(10-17)-(-25)]-1=$
    $[91-(-7)+25]-1$
    $[91+7+25]-1$
    $[123]-1$
    122

15. $-4^2+12\div3=$
    $-16+4$
    -12

7. $12-(-6)^2+2\cdot5=$
   $12-(36)+10$
   $-24+10$
   -14

9. $60\cdot2\div4\cdot8=$

   $120\div4\cdot8$

   $30\cdot8$

   240

17. $1-6|4-7|+3^2=$
   $1-6|-3|+9$
   $1-6\cdot3+9$
   $1-18+9$
   -8

19. $\dfrac{15-42\div7\bullet3}{4+3^3\div(-9)}=$
   $\dfrac{15-6\bullet3}{4+27\div(-9)}$
   $\dfrac{15-18}{4-3}$
   $\dfrac{-3}{1}$
   -3

Homework

1. dwell time = 4 hours (the time a crew member is on a plane)
   vacant time = 2 hours (the time the plane is not in flight)
   free time = 6 hours (the time the crew is not on the plane)
   number of hands = 3 (the number of planes)
   number of balls = unknown (the number of crews)
   period of "hands" = (dwell time) + (vacant time)
   = 4 hours + 2 hours
   = 6 hours
   period of "balls" = (dwell time) + (free time)
   = 4 hours + 6 hours
   = 10 hours

   The juggling proportion is:

   $$\frac{\text{period of hands}}{\text{period of balls}}=\frac{\text{number of hands}}{\text{number of balls}}$$

   $\dfrac{6}{10}=\dfrac{3}{x}$ where x represents the unknown number of balls (crews)

   $6x=3\cdot10$ cross multiply to form an equation without fractions

   $x=5$ divide both sides of the equation by 6 to find x.

   There will need to be 5 crews, assuming they can work around the clock and are in the correct location to service a flight.

2. dwell time = 4.5 hours (the time a widget machine is on-line, with one operator)
   vacant time = 0.5 hours (the time the widget machine is off-line)
   free time = x hours (the time the employee is not operating the widget machine)
   number of hands = 5 (the number of widget machines)
   number of balls = 7 (the number of employees)
   period of "hands" = (dwell time) + (vacant time)
   = 4.5 hours + 0.5 hours
   = 5 hours
   period of "balls" = (dwell time) + (free time)
   = 4.5 hours + x hours
   = $4.5+x$

The juggling proportion is:

$$\frac{\text{period of hands}}{\text{period of balls}} = \frac{\text{number of hands}}{\text{number of balls}}$$

$\frac{5}{4.5+x} = \frac{5}{7}$ where x represents the unknown free time

$7 \cdot 5 = 5(4.5 + x)$ cross multiply to form an equation without fractions

$35 = 22.5 + 5x$ multiply

$12.5 = 5x$ subtract 22.5 from both sides of the equation

$2.5 = x$ divide by 5

Employees will get 2.5 hours free between shifts on the widget machine.

3. dwell time = 30 minutes (the time a theater is being cleaned)
vacant time = 120 minutes (the time the theater can't be cleaned)
free time = x minutes (the time the cleaning crew isn't cleaning)
number of hands = 10 (the number of theaters)
number of balls = 3 (the number of cleaning crews)
period of "hands" = (dwell time) + (vacant time)
= 30 minutes + 120 minutes
= 150 minutes
period of "balls" = (dwell time) + (free time)
= 30 minutes + x
= $30 + x$

The juggling proportion is:

$$\frac{\text{period of hands}}{\text{period of balls}} = \frac{\text{number of hands}}{\text{number of balls}}$$

$\frac{150}{30+x} = \frac{10}{3}$ where x represents the unknown free time

$150 \cdot 3 = 10(30 + x)$ cross multiply to form an equation without fractions

$450 = 300 + 10x$ multiply

$150 = 10x$ subtract 300 from both sides of the equation

$15 = x$ divide by 10

Cleaning crews will get 15 minutes free.

5. a. $\frac{1}{R} = \frac{r_1 + r_2}{r_1 r_2}$

$\frac{1}{R} = \frac{5+5}{5 \cdot 5}$ substitute resistance values into the formula

$\frac{1}{R} = \frac{10}{25}$ simplify

$25 = 10R$ cross multiply

$2.5 = R$ solve

The total resistance is 2.5 ohms.

b. $\frac{1}{R} = \frac{r_1 + r_2}{r_1 r_2}$

$\frac{1}{R} = \frac{8+4}{8 \cdot 4}$ substitute resistance values into the formula

$\frac{1}{R} = \frac{12}{32}$ simplify

$32 = 12R$ cross multiply

$2.\overline{6} = R$ solve

The total resistance is $2.\overline{6}$ ohms.

6. b. $\frac{C}{D} = \frac{22}{7}$

$\frac{25}{D} = \frac{22}{7}$ substitute value of circumference into the formula

$25 \cdot 7 = 22 \cdot D$ cross multiply

$175 = 22D$ multiply

$7.95 \approx D$ divide by 22

The diameter, D, is 8.0 inches, to the nearest tenth.

c. $\frac{C}{D} = \frac{22}{7}$

$\frac{C}{15} = \frac{22}{7}$ substitute value of circumference into the formula

$7 \cdot C = 22 \cdot 15$ cross multiply

$7C = 330$ multiply

$C \approx 47.14$ divide by 7

The circumference, C, is 47.14 cm, to the nearest hundredth.

Lesson 9 Fractions

Homework

2. c. $\frac{4}{9} = 0.\overline{4}$, 0.4 is similar, but different. d. $\frac{4}{11} = 0.\overline{36}$, 0.36 is similar but different

e. $\frac{188}{333} = 0.\overline{564}$, 0.564 is similar but different

f. $\frac{2}{7} = 0.\overline{285714}$, 0.285714 is similar but different

g. $\frac{3}{7} = 0.\overline{428571}$, 0.428571 is similar but different

3. a. Dennis bought 2 pounds + 6 pounds = 8 pounds of nuts.

b. The ratio of cashews to nuts $= \frac{\text{parts of cashews}}{\text{parts of nuts}}$

$= \frac{2}{8}$

$= \frac{1}{4}$

5. When several fractions have the same numerator, a fraction with a smaller denominator is larger. Therefore, the fractions must be arranged in ascending order (from smallest to largest) of the denominators.

$$\frac{5}{2}, \frac{5}{6}, \frac{5}{7}, \frac{5}{12}, \frac{5}{21}$$

7.

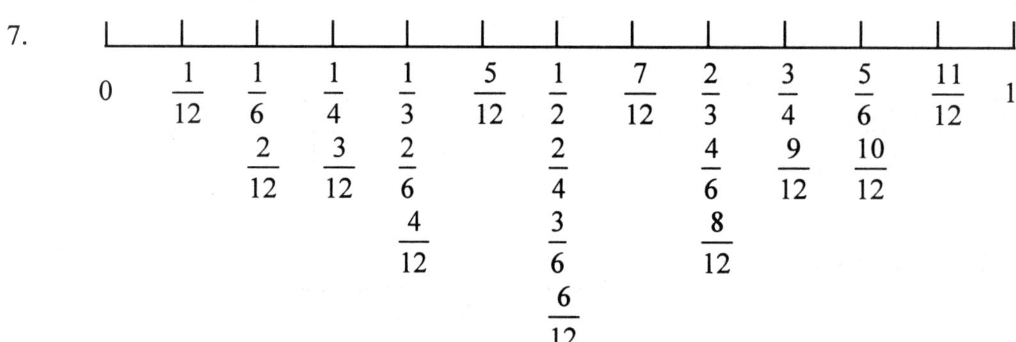

b. The following fractions are equivalent, as seen on the line:

$$\frac{1}{6}=\frac{2}{12}, \frac{1}{4}=\frac{3}{12}, \frac{1}{3}=\frac{2}{6}=\frac{4}{12}, \frac{1}{2}=\frac{2}{4}=\frac{3}{6}=\frac{6}{12}$$

$$\frac{2}{3}=\frac{4}{6}=\frac{8}{12}, \frac{3}{4}=\frac{9}{12}, \frac{5}{6}=\frac{10}{12}$$

8. Since all the numerators are equal, N, the smaller the denominator the larger the fraction.

$$\frac{N}{2} \quad \frac{N}{3} \quad \frac{N}{5} \quad \frac{N}{8} \quad \frac{N}{9}$$

9. Since all the denominators are equal, D, the larger the numerator, the larger the fraction.

$$\frac{7}{D} \quad \frac{6}{D} \quad \frac{5}{D} \quad \frac{4}{D} \quad \frac{2}{D} \quad \frac{1}{D}$$

Lesson 10 Unit Analysis

Homework

2. Prefixes used in the metric system are: milli, centi, deci, no prefix, deka, hecto, kilo

4. a. $\frac{26 \text{ miles}}{1} \cdot \frac{1.609 \text{ km}}{1 \text{ mile}} \approx 41.834$ kilometers

b. $\frac{75 \text{ liters}}{1} \cdot \frac{1 \text{ gallon}}{3.785 \text{ liters}} \approx 19.815$ gallons

5. a. $\frac{5 \text{ feet} \cdot \text{feet}}{1} \cdot \frac{12 \text{ inches}}{1 \text{ foot}} \cdot \frac{12 \text{ inches}}{1 \text{ foot}} = 720$ square inches

b. $\frac{134 \text{ inches} \cdot \text{inches}}{1} \cdot \frac{1 \text{ foot}}{12 \text{ inchest}} \cdot \frac{1 \text{ foot}}{12 \text{ inches}} = 0.93$ sq. ft.

6. a. $\frac{1 \text{ foot} \cdot \text{foot} \cdot \text{foot}}{1} \cdot \frac{12 \text{ inches}}{1 \text{ foot}} \cdot \frac{12 \text{ inches}}{1 \text{ foot}} \cdot \frac{12 \text{ inches}}{1 \text{ foot}} = 1728$ cu. in.

b. $\frac{2160 \text{ inches} \cdot \text{inches} \cdot \text{inches}}{1} \cdot \frac{1 \text{ foot}}{12 \text{ inches}} \cdot \frac{1 \text{ foot}}{12 \text{ inches}} \cdot \frac{1 \text{ foot}}{12 \text{ inches}} = 1.25$ cu. ft.

8. $\frac{68 \text{ miles}}{1 \text{ hour}} \cdot \frac{1 \text{ hour}}{60 \text{ min}} \cdot \frac{1 \text{ min}}{60 \text{ sec}} \cdot \frac{5280 \text{ feet}}{1 \text{ mile}} \cdot \frac{12 \text{ inches}}{1 \text{ foot}} \approx 1196.8$ inches per second

11. a. $\frac{75 \text{ newton} \cdot \text{meter}}{1 \text{ sec}} \cdot \frac{1 \text{ pound}}{4.448 \text{ newtons}} \cdot \frac{3 \text{ feet}}{0.9114 \text{ meters}} \approx 55.32$ foot pounds per second

$\frac{55.32 \text{ foot} \cdot \text{pound}}{1 \text{ sec}} \cdot \frac{1 \text{ horsepower}}{\frac{550 \text{ foot} \cdot \text{pound}}{\text{sec}}} \approx 0.1$ horsepower.

b. $\frac{1 \text{ bar}}{1} \cdot \frac{100{,}000 \text{ newtons}}{\text{m} \cdot \text{m}} \cdot \frac{1 \text{ pound}}{4.448 \text{ newtons}} \cdot \frac{0.3048 \text{ m}}{12 \text{ inches}} \cdot \frac{0.3048 \text{ m}}{12 \text{ inches}} \approx 14.5$ psi.

Lesson 11 Work and Adding Fractions

Homework

3. Mike's rate + John's rate + Judy's rate = combined work rate

$4 + 3 + 6 = 13$ cars per hour

$\frac{13 \text{ cars}}{60 \text{ minutes}}$ is $\frac{60 \text{ minutes}}{13 \text{ cars}}$ or about 4.615 minutes per car

Data and Graphs

Lesson 1 May I Take Your Order?

Homework

4. It looks like revenue is going up during the years shown. We do not know whose revenue it is, or what the units the revenue is in. Is it in dollars, or millions of lira or what?

Lesson 2 Would You Like Pie With That?

The Nitty Gritty – Conversion Among Percentage, Fraction, and Decimal Forms

Percent	Fraction	Decimal
25%	$\frac{1}{4}$	0.25
30%	3/10	0.3
120%	6/5	1.2
200%	2	2.0
266.6…%	$\frac{8}{3}$	2.6666…
33.3…%	1/3	0.333333...
20%	$\frac{1}{5}$	0.2
16%	4/25	0.16
166.6…%	5/3	1.666…
350%	7/2	3.5
83.3…%	$\frac{5}{6}$	0.8333…
75%	3/4	0.75
100%	1/1	1.0

Homework

1. a. Pizza Hut revenue outside the U.S. was about $700,000,000 in 1989.
   b. The revenue in 1991 might be the average of the revenue in 1990 and 1992.

$$\frac{\text{Revenue in 1990} + \text{Revenue in 1992}}{2} = \frac{800 + 960}{2} = \$880 \text{ million}$$

   c. The revenue is growing at about $80 million per year, based on 1990 to 1992. The 1993 revenue would be about $80 more than the 1992 revenue
   80 + 960 = 1040 million, or $1,040,000,000.

2. a. We know the number of Burger King restaurants to be 10,000 which is 28% of the total.
   0.28(Total number of restaurants worldwide in 1992) = 10,000

$$\text{Total number of restaurants world wide in 1992} = \frac{10{,}000}{0.28} \approx 35{,}714 \text{ restaurants.}$$

   McDonald's has 55% of the total or 0.55(35,714) ≈ 19,643 restaurants.
   Taco Bell has 17% of the total or 0.17(35,714) ≈ 6071 restaurants.

   Another approach to this problem is to think in terms of proportions. Then we don't need to calculate the total number of restaurants worldwide.

$$\frac{\#\text{ of Burger Kings}}{0.28} = \frac{\#\text{ of McDonald's}}{0.55} = \frac{\#\text{ of Taco Bells}}{0.17}$$

$$\frac{10{,}000}{0.28} = \frac{\#\text{ of McDonald's}}{0.55} \text{ or } \#\text{ of McDonald's} = \frac{0.55 \cdot 10{,}000}{0.28} \approx 19{,}643.$$

$$\frac{10{,}000}{0.28} = \frac{\#\text{ of Taco Bells}}{0.17} \text{ or } \#\text{ of Taco Bells} = \frac{0.17 \cdot 10{,}000}{0.28} \approx 6071$$

   b. Using proportions,

$$\frac{\#\text{ of Burger Kings}}{0.28} = \frac{\#\text{ of McDonald's}}{0.55} = \frac{\#\text{ of Taco Bells}}{0.17}$$

$$\frac{8000}{0.17} = \frac{\#\text{ of McDonald's}}{0.55} \text{ or } \#\text{ of McDonald's} = \frac{0.55 \cdot 8000}{0.17} \approx 25{,}882.$$

$$\frac{8000}{0.17} = \frac{\#\text{ of Burger Kings}}{0.28} \text{ or } \#\text{ of Burger Kings} = \frac{0.28 \cdot 8000}{0.17} \approx 13{,}176.$$

Lesson 3 The Graphing Gourmet

Homework

1.

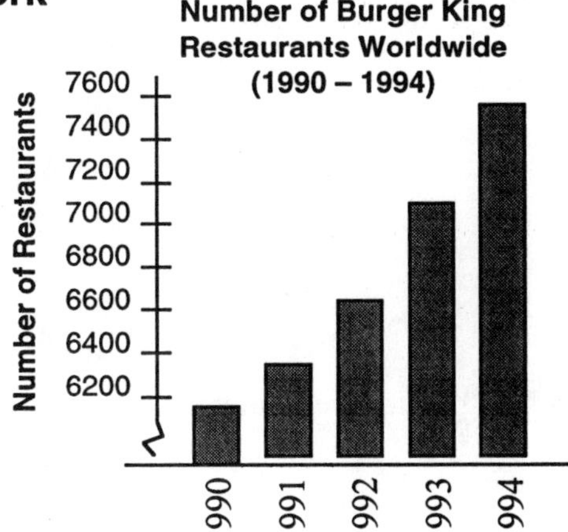

2. Burger King $\frac{7547}{50,280} \cdot 360^\circ \approx 54^\circ$

Sbarro $\frac{729}{50,280} \cdot 360^\circ \approx 5^\circ$

Pizza Hut $\frac{11,546}{50,280} \cdot 360^\circ \approx 83^\circ$

Taco Bell $\frac{5846}{50,280} \cdot 360^\circ \approx 42^\circ$

KFC $\frac{9407}{50,280} \cdot 360^\circ \approx 67^\circ$

McDonald's $\frac{15,205}{50,280} \cdot 360^\circ \approx 109^\circ$

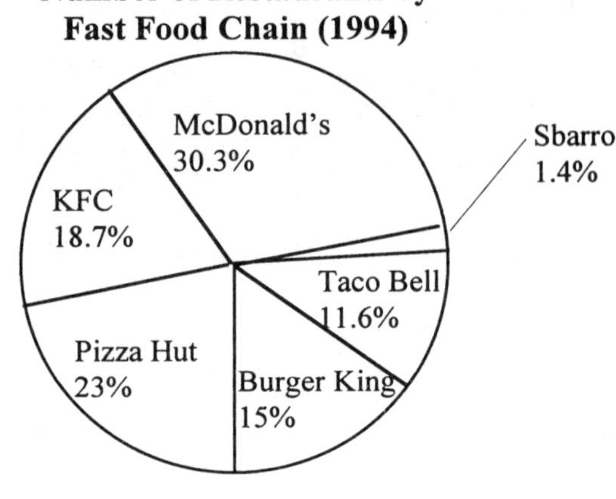

6.

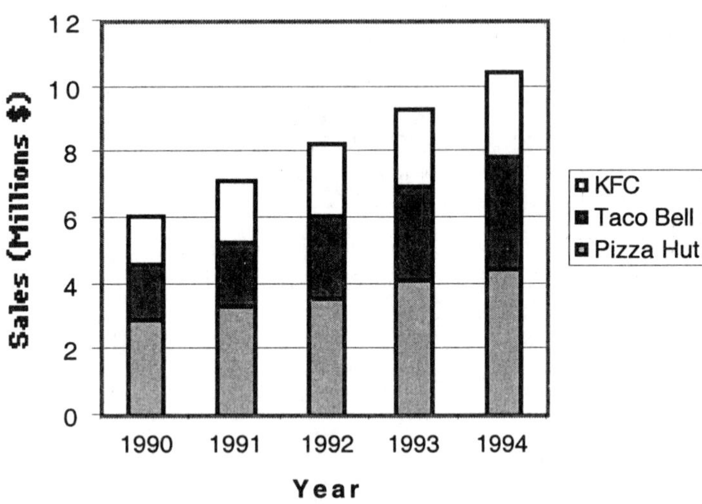

Lesson 4 Waiter, There's a Whisker in My Soup

The Nitty Gritty - Percent Increase

Problems

1. a. $6.9 - 4.1 = \$2.8$ billion b. $\frac{2.8}{4.1} \approx 68\%$

3. a. $7.1 - 5.4 = \$1.7$ billion b. $\frac{1.7}{5.4} \approx 31\%$

4. Pizza Hut had the greatest growth, but Taco Bell had the greatest percent growth.

Homework

1. a. Based on the table for Worldwide System Sales 1989 – 1994, Pizza Hut has a mean of $5.55 billion, a median of $5.5 billion and no mode.

Mean: $\frac{4.1+4.9+5.3+5.7+6.4+6.9}{6} \approx \$5.55$ billion

Median: $\frac{5.3+5.7}{2} \approx \$5.5$ billion

Taco Bell has a mean of $3.17 billion, a median of $3.05 billion and no mode.

Mean: $\frac{2.1+2.4+2.8+3.3+3.9+4.5}{6} \approx \$3.17$ billion

Median: $\frac{2.8+3.3}{2} = \$3.05$ billion

KFC has a mean of $6.38 billion, a median of $6.45 billion and a mode of $7.1 billion.

Mean: $\frac{5.4+5.8+6.2+6.7+7.1+7.1}{6} \approx \$6.38$ billion

Median: $\frac{6.2+6.7}{2} \approx \$6.45$ billion

b. Mode does not apply to the data for Pizza Hut and Taco Bell. For KFC, mode is not a very good measure of central tendency since it is the largest value. With this data the mode describes when the worldwide system sales stayed the same, which is not necessarily related to "central" tendency.

c. In 1994, Pizza Hut had 11,546 restaurants worldwide and made 6.9 billion dollars in sales. In 1994 Taco Bell had 5,846 restaurants worldwide and made 4.5 billion dollars in sales. In 1994 KFC had 9,407 restaurants worldwide and made 7.1 billion dollars in sales.

d. Pizza Hut made $\frac{6{,}900{,}000{,}000}{11{,}546} \approx \$597{,}610$ per restaurant in 1994.

Taco Bell made $\frac{4{,}500{,}000{,}000}{5846} \approx \$769{,}757$ per restaurant in 1994.

KFC made $\frac{7{,}100{,}000{,}000}{9407} \approx \$754{,}757$ per restaurant in 1994.

2. a. mean: $\frac{20{,}000+20{,}000+22{,}000+23{,}000+25{,}000+26{,}000+27{,}000+34{,}000+1{,}234{,}000}{9}$

$= \frac{1{,}431{,}000}{9} = \$159{,}000$ per player.

Median: $25,000
Mode: $20,000

Exponential Growth and Decay

Lesson 1 Patterns of Growth

Homework

1. a.

Town A

Year	Pop.	Diff.	Ratio
1980	1250		
		252	1.20
1985	1502		
		249	1.17
1990	1751		
		247	1.14
1995	1998		

b.

Town B

Year	Pop.	Diff.	Ratio
1980	2734		
		−154	0.94
1985	2580		
		−147	0.94
1990	2433		
		−138	0.94
1995	2295		

c.

Town C

Year	Pop.	Diff.	Ratio
1980	1545		
		310	1.20
1985	1855		
		370	1.20
1990	2225		
		445	1.20
1995	2370		

Town A exhibits linear, or arithmetic, growth at the rate of about 250 people every 5 years. Town B exhibits nearly linear decay at the rate of about 147 people every 5 years and also exponential decay with a decay factor of 0.94 every 5 years. Town C exhibits exponential growth with growth factor of 1.20 every 5 years.

3.

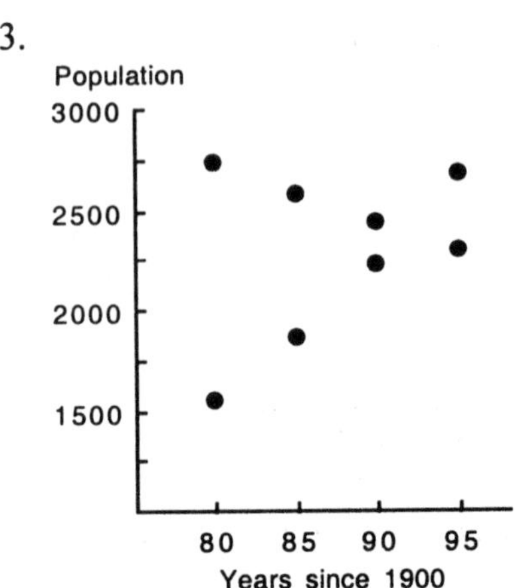

a. Town B's population looks linear and is decreasing. Town C's population appears to curve upward and is increasing.
b. The two graphs seem to cross at (92, 2400).
c. The intersection is when both towns have the same population of 2400 people in 1992.

7. a. difference = 260, next two terms = 1060, 1320
b. difference = −6, next two terms = 51, 45
c. difference = 3.8, next two terms = 36.9, 40.7

8. a. ratio = 1.5, next two terms = 1575, 2363
   b. ratio = 1.1, next two terms = 49.5, 54.45
   c. ratio = 0.75, next two terms = 31.5, 23.625

9. a. If d = 1.5, the sequence is increasing.
   b. If d = –40, the sequence is decreasing.
   c. If d = –0.85, the sequence is decreasing.

Lesson 2 Representing Growth

Homework

1.

Year	n	Wages	
1977	0	26.93	difference
1978	1	28.84	1.91
1979	2	31.16	2.32
1980	3	34.42	3.26
1981	4	37.27	2.85
1982	5	40.00	2.73
1983	6	42.82	2.82

Table 6

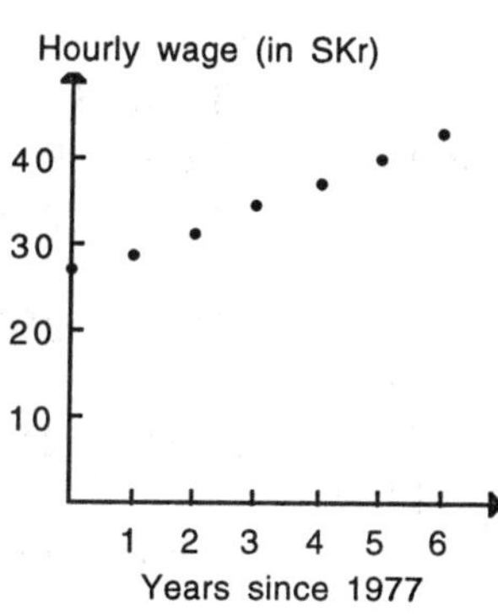

Verbal: The average growth in the hourly wage is 2.65 SKr per year, starting at 26.93.
Formula: $w(n) = 26.93 + 2.65n$

The slope of the linear equation represents the average yearly difference, The linear interpretation gives an average growth but ignores that the rate varies from 1.91 to 3.26 Swedish Kroner.

3.

Year	n	Wages	
1955	0	3.446	ratio
1960	5	3.595	1.043
1965	10	3.738	1.040
1970	15	3.888	1.040
1975	20	4.017	1.033

Table 8

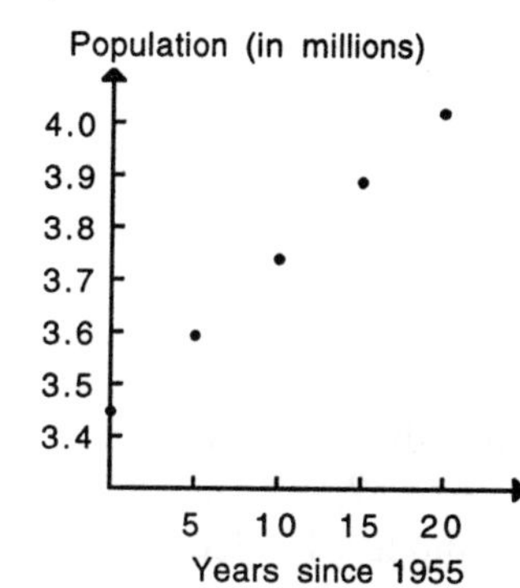

Verbal: The average growth rate for Norway is 1.039 for each 5-year period from 1955 to 1975.
Formula: $p(n) = 3.446(1.039)^n$

The base 1.039 represents the growth rate in a 5-year period. The exponent n represents the number of 5-year periods, where n refers to the starting period.

5.

Year	1987	1988	1989	1990	1991
Lead ($\mu g/m^3$)	0.106	0.088	0.072	0.062	0.053

ratios: 0.83 0.82 0.86 0.85

The concentration of lead in the air is decreasing. The common ratio of about 0.84 shows that the concentration is decreasing exponentially.

Formula: $C(n) = 0.106 \cdot (0.84)^n$, n in years since 1987

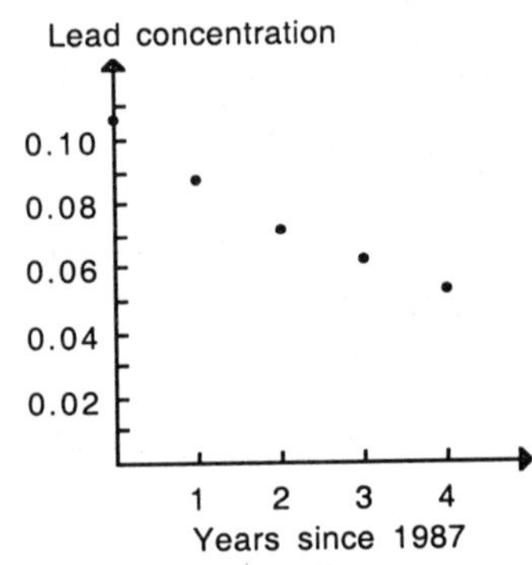

6. a. Anytown started with a population of 10,000 and has decreased by 150 people per year. It is declining by a fixed number of people each year.

b.

Year	0	1	2	3
Population	10,000	9,850	9,700	9,550

9. a. $d(n) = 900(1.21)^n$

b.

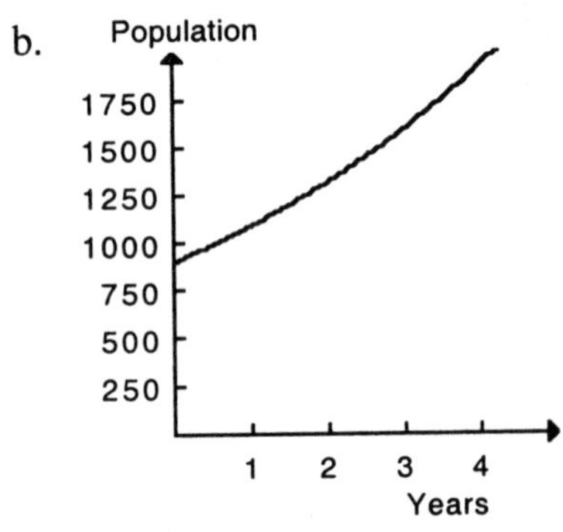

11. a. In about year 28, the investment will be worth $200,000.
    b. After 28 years, the 10% investment is worth $125,000.
    c. $650,008.67 – $452,592.56 = $197,416.11
    d. A seemingly modest 1% fee can mean many dollars unearned over a period of 40 years.

Lesson 3 Doubling

Homework

1. $\text{sal}(n) = 3 \cdot 2^{n-1}$, n in weeks

Weeks, n	Salary $= 3 \cdot 2^{n-1}$
1	$3 \cdot 2^{1-1} = 3$
2	$3 \cdot 2^{2-1} = 6$
3	$3 \cdot 2^{3-1} = 12$
4	$3 \cdot 2^{4-1} = 24$

3. One less doubling, so at age 55, she would be making half her salary.

5. a.

Months	0	1	2	3	4	5
Shoes	20	40	80	160	320	340

$S(n) = 20 \cdot 2^n$, where n is the number of months.

b. Starting with 20 shoes, the number of shoes sold has doubled each month for 5 months.

8.

Time	0	1	2	3	4	5
# Bacteria	600	1200	2400	4800	9600	19,200

a. The bacteria population will double in less than 5 hours.

b. 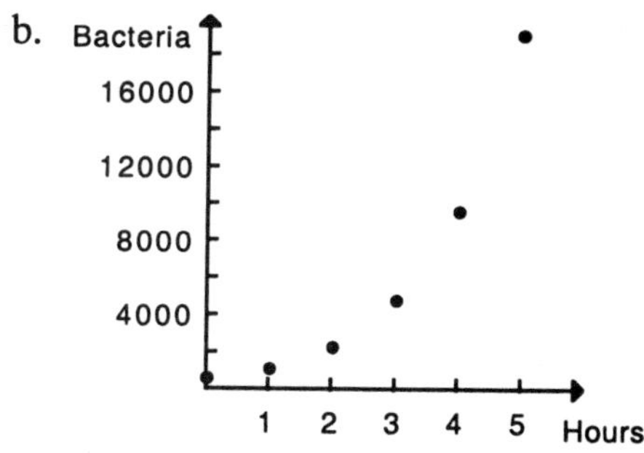

c. $p(n) = 600 \cdot 2^n$, where n is the number of hours.

10.

Day, n	# Bacteria
0	9,000,000
1	4,500,000
2	2,250,000
3	1,125,000
4	562,000
5	281,250
6	140,625
7	70,312
8	35,156
9	17,578
10	8,789
11	4,394
12	2,197
13	1,098
14	549

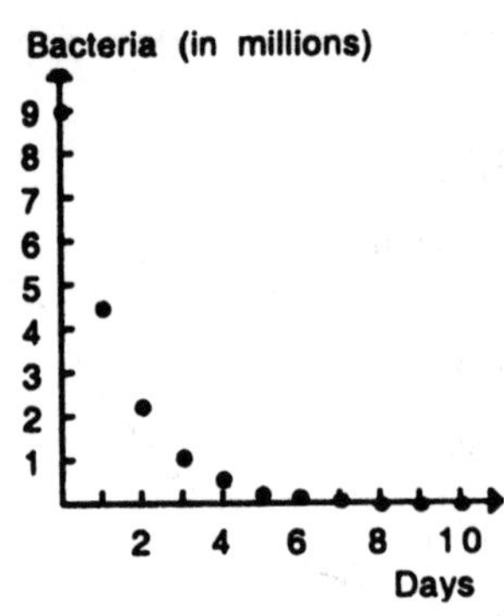

The number of bacteria will be below 1000 after 14 days.

Lesson 4 Doubling Variations

Homework

1. a. 5 months

   b. $0.25n = n/4$, 4 days

   c. $0.3n = \dfrac{n}{\left(\frac{10}{3}\right)}$, 10/3 days

   d. $3n = \dfrac{n}{\left(\frac{1}{3}\right)}$, 1/3 of a week

   e. 0.05 years

   f. 7 years

2. The doubling time is 20 hours for the first and 20 months for the second.

3. Formula: $B = 1000 \cdot 2^{\frac{n}{12}}$

Table:

Hours	# Bacteria
0	1000
1	1059
2	1122
3	1189
4	1260
5	1335
6	1414
7	1498
8	1587
9	1682
10	1782
11	1888
12	2000

Graph:

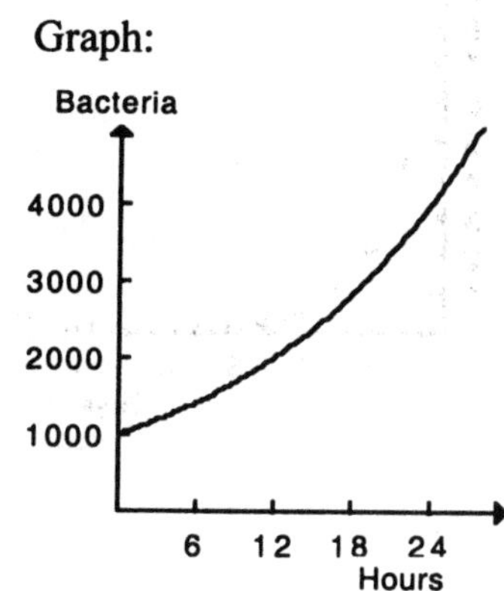

5. Verbal: Something starts at 0.321 and doubles every 1.43 time periods.

Table:

Time Period	Amount
0	0.321
1	0.521
2	0.847
3	1.376
4	2.236
5	3.632
6	5.900
7	9.584
8	15.569
9	25.293
10	41.088

Graph:

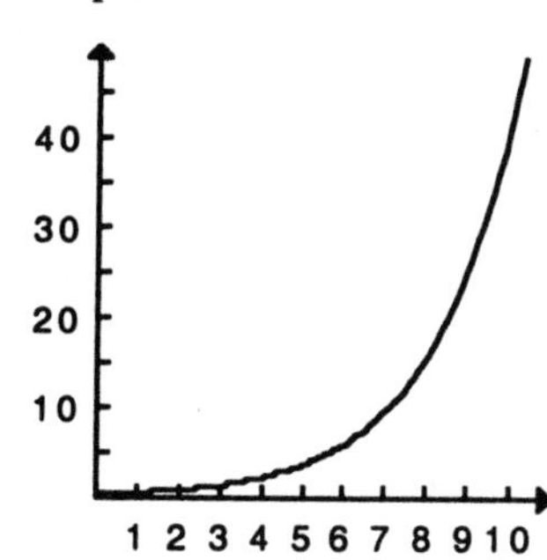

7. Formula: $S = 4.5(1/2)^{n/2400}$, n in years

Table:

Years	Amount (kg)
0	4.500
200	4.247
400	4.009
600	3.784
800	3.572
1000	3.371
1200	3.182
1400	3.003
1600	2.835
1800	2.676
2000	2.526
2200	2.384
2400	2.250

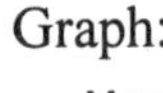
Graph:

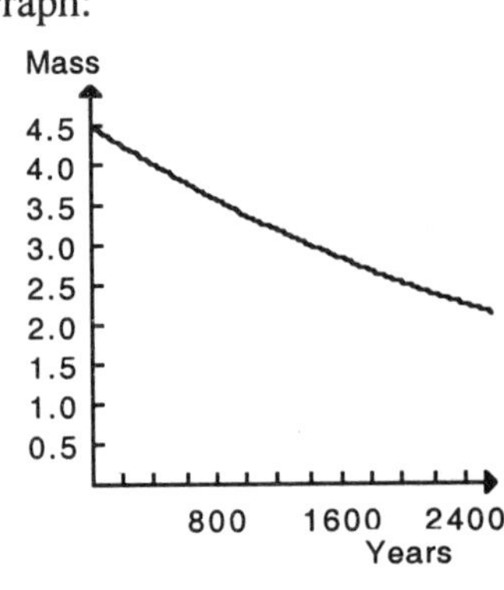

8. a. 3^7 b. 5^9 c. b^{11} d. 1 e. 2^{-4} f. $(1/2)^4$ g. 2^{21} h. 3^{-8} i. $2^{0.3}$ j. x^3 k. 2^6 l. 3^3

Lesson 5 Beyond Doubling

Homework

1. $1500*3^{(x/9)}$ It triples every 9 hours.

2.

Year	Population
5	$\sqrt{5155 \cdot 7007} \approx 6010$
15	$\sqrt{7007 \cdot 9526} \approx 8170$
25	$\sqrt{9526 \cdot 12{,}949} \approx 11{,}106$
35	$\sqrt{12{,}949 \cdot 17{,}602} \approx 15{,}097$

6. As b gets close to 1, the curve flattens out to a horizontal line.

x	1.1^x	1.01^x	0.99^x	0.9^x
–4	0.683	0.961	1.04	1.52
–3	0.751	0.971	1.03	1.37
–2	0.826	0.980	1.02	1.23
–1	0.909	0.990	1.01	1.11
0	1	1	1	1
1	1.1	1.01	0.99	0.9
2	1.21	1.02	0.980	0.81
3	1.331	1.03	0.970	0.729
4	1.4641	1.04	0.961	0.6561

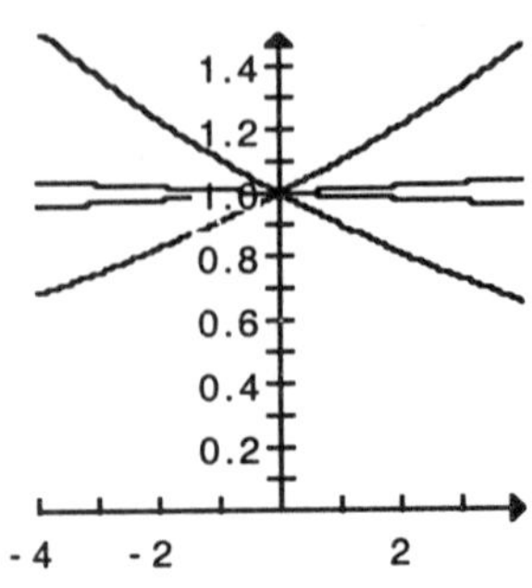

Lesson 6 Relating Different Bases

Homework

1. By the Rule of 72, the doubling time is about 72/12 = 6. Begin the test and refine method with $x = 6$. $1.12^6 \approx 1.97$. Try $x = 6.1$, $1.12^{6.1} \approx 1.996$. Try $x = 6.11$, $1.12^{6.11} \approx 1.9986$. Try $x = 6.12$, $1.12^{6.12} \approx 2.008$. Try $x = 6.115$, $1.12^{6.115} \approx 1.9997$. Try $x = 6.116$, $1.12^{6.116} \approx 1.9999$. Try $x = 6.117$, $1.12^{6.117} \approx 2.0002$. Use $x = 6.116$. The doubling formula is $1.12^x = 2^{x/6.116}$. Something that doubles in 6.116 years has a 12% annual growth rate.

x	0	1	2	3	4	5
1.12^x	1	1.12	1.25	1.40	1.57	1.76
$2^{x/6.116}$	1	1.12	1.25	1.40	1.57	1.76

2. $2^{x/15} = (2^{1/15})^x \approx 1.0473^x$. This is 4.73% growth. Something that doubles every 15 hours has a 4.73% hourly growth rate.

x	0	1	2	3	4	5
1.047^x	1	1.047	1.096	1.148	1.202	1.258
$2^{x/15}$	1	1.047	1.097	1.149	1.203	1.260

6. Solve $e^{0.08x} = 2$

x	9	8.5	8.6	8.65	8.66	8.67
$e^{0.08x}$	2.05	1.97	1.9897	1.9977	1.9993	2.0009

The doubling period is 8.66, so $e^{0.08x} \approx 2^{x/8.66}$.

8. $A = 1250 \cdot 2^{n/6}$, n in years

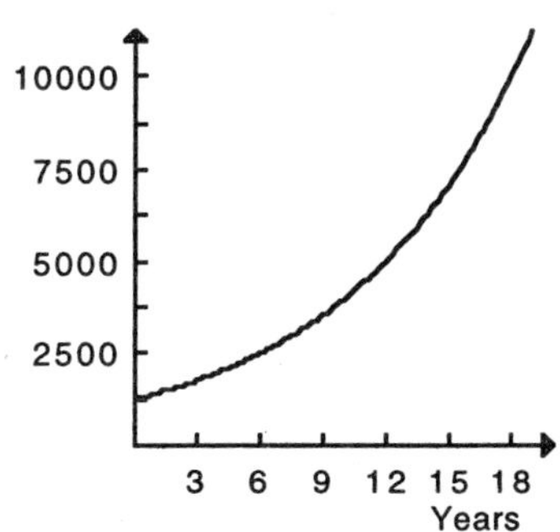

11. $A = 15(1/2)^{n/35,000}$, n in years

n	10,000	45,000	80,000	115,000
$15(1/2)^{n/35,000}$	12.3050	6.1525	3.0763	1.5381

Note that halving occurs every 35,000 years in the table.

13. An initial amount of 20 grows at a rate of 9% every time period.

16. Something grows from 300 at zero and grows to 600 in 5 time periods. It doubles every 5 time periods. Formula: $y = 300(2)^{x/5}$.

18. It decays at 4.52% per year. $y = (0.9548)^n$.

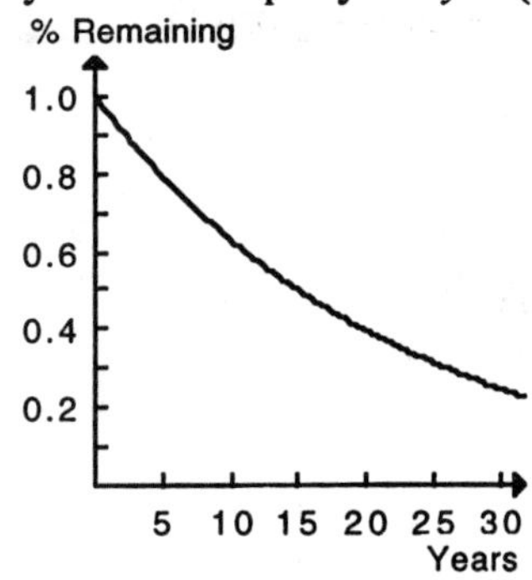

Lesson 7 Overpopulation

Homework

1. Answers will vary. In 1975, using the geometric mean, the population was about 4.06 billion.

5. $P = 2.53 \cdot (1.0186)^x$

10. $(1.48)^{1/8} \approx 1.05$. So it rose by 5% per year, on average.

Lesson 8–A Look at Logs

Homework

2. By the Rule of 72, the doubling time is about $72/11 \approx 6.5$
   The doubling time is the solution to the equation:
   $1.11^x = 2$
   $\log_{1.11} 2 = x$
   $x = \dfrac{\log 2}{\log 1.11} \approx 6.6419$

3. a. $3^5 = x + 4$ b. $7^{-2} = 2x^2$ c. $x^3 = 35$ d. $x^z = y$

4. a. $\log_{25} x = 2$ b. $\log_x 625 = 4$ c. $\log_3 78 = x$ d. $\log_a c = b$

9. a. $\log a + \log b$ b. $\ln a + \ln b + \ln c$ c. $\log a - \log b$
   d. $\log a + \log c - \log b$ e. $\log a - \log b - \log c$ f. $3 \cdot \log a$
   g. $3(\ln a + \ln b)$ h. $2(\log a - \log b)$ i. $n \cdot (\log a + \log c - \log b)$

10. a. $\log (xy)$ b. $\ln (x/2y)$ c. $\log (xy)^2$
    d. $\log (xy/z)^4$ e. $\log (x^3/y)$ f. $\ln (1/2)$

11. a. Solve $10^{0.023x} = 2^{nx}$, for n.
$10^{0.023} = 2^n$
$n = \log_2 10^{0.023}$
$n = 0.023\dfrac{\log 10}{\log 2}$
$n \approx 0.0764$
$10^{0.023x} \approx 2^{0.0764x}$
The doubling time is 13.088.

$10^{0.023x} = (10^{0.023})^x = 1.0544^x$
Find the doubling time for 5.44% growth
$1.0544^t = 2$
$t = \log_{1.0544} 2$
$t = \dfrac{\log 2}{\log 1.0544} \approx 13.088$
$10^{0.023x} \approx 2^{x/13.088} = 2^{0.0764x}$

b. Similarly to part a, $10^{0.23x} \approx 2^{0.764x}$
or
Using the result from part a, $10^{0.23x} = (10^{0.023x})^{10} \approx (2^{0.0764x})^{10} = 2^{0.764x}$
The doubling time is 1.3088.

c. Using the result from part b, $10^{2.3x} = (10^{0.23x})^{10} \approx (2^{0.764x})^{10} = 2^{7.64x}$
The doubling time is 0.13088

d. Similarly to part a, $e^{0.05x} \approx 2^{0.0721x}$
The doubling time is 13.87.

e. $e^{0.25x} \approx 2^{0.361x}$
The doubling time is 2.77.

f. $e^{0.35x} \approx 2^{0.505x}$
The doubling time is 1.98.

14. a. $1.07^x = 2$
$\log 1.07^x = \log 2$
$x \log 1.07 = \log 2$
$x = (\log 2)/(\log 1.07)$
$x \approx 10.245$ years

b. $1.012^x = 2$
$\log 1.012^x = \log 2$
$x \log 1.012 = \log 2$
$x = (\log 2)/(\log 1.012)$
$x \approx 58.1$ years

15. a. Hourly growth factor $= 2^{1/14} \approx 1.0508$. The percent growth rate is 5.08%.

b. Annual growth factor $= 3^{1/18} \approx 1.0629$. Now solve $1.0629^t = 2$, to get $t = 11.36$ years

16. a. $3^{x-3} = 100$
$x - 3 = \log_3 100$
$x = 3 + \dfrac{\log 100}{\log 3}$
$x \approx 7.192$

b. $100e^{-0.02x} = 50$
$e^{-0.02x} = 0.5$
$-0.02x = \ln 0.5$
$x = -(\ln 0.5)/0.02$
$x \approx 34.657$

c. $\ln x = 9.15$
$x = e^{9.15}$
$x \approx 9414.440$

d. $\log 3x = 1.75$
$3x = 10^{1.75}$
$x = 10^{1.75}/3$
$x \approx 18.745$

e. $\log_5 (x + 1) = 2.5$
$x + 1 = 5^{2.5}$
$x = 5^{2.5} - 1$
$x \approx 54.902$

f. $3 \cdot 10^{2x} = 15$
$10^{2x} = 15/3 = 5$
$2x = \log 5$
$x = (\log 5)/2$
$x \approx 0.346$

Lesson 9 A Sense for Logarithms

Homework

4. 10 more decibels.

6. A decrease of 3 decibels corresponds to halving the acoustic power. The second decrease of 3 halves it again. (0.5)(0.5) = 0.25. The result is one fourth the acoustic power.

8. a. $n = 10 \cdot \log P_1 - 10 \cdot \log P_0$ b. $n = 10 \cdot \log 2 + 10 \cdot \log P_1 - 10 \cdot \log P_0$

10. One voice is 62 dB. 2 is 65 dB. 4 is 68 dB. 8 is 71 dB. and 16 is 74 dB. So 15 people would be just under 74 dB.

11. $10 \cdot \log((0.32 + 0.32)/1 \times 10^{-12})$ or 118 dB.

15. (92 + 0.8 + 0.6) = 93.4 dB. It's close. Remember to start with the loudest sound.

Lesson 10 Life in a Petri Dish

Homework

1. Semi-log graph:

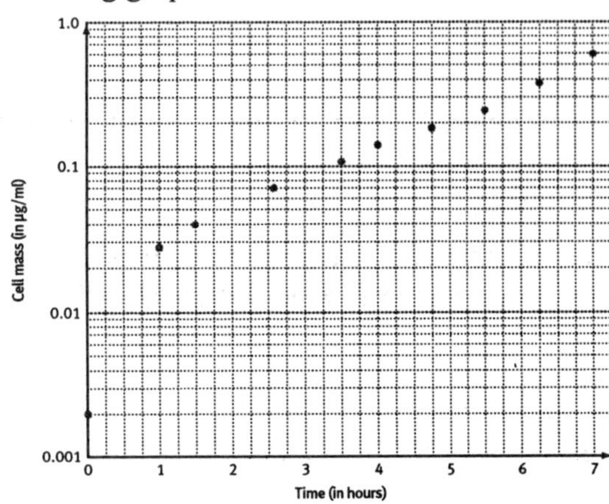

4. $y = 0.6535x - 4.8$

11. It is exponential decay, really. Since the log of the data looks linear, it demonstrates that the data is exponential.

13. a. $D = 8.058(0.8554)^G$
    b. The middle part of the data looks exponential, that is the part of the graph that looks linear.

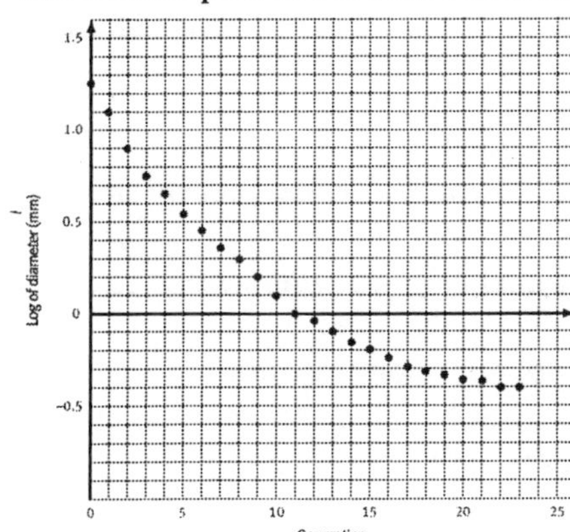

14. a. $10^x = 13.8$
    $\log 13.8 = x$
    $x \approx 1.14$

    b. $10^{3x} = 24$
    $\log 24 = 3x$
    $x = (\log 24)/3$
    $x \approx 0.46$

    c. $10^{3+x} = 19.5$
    $\log 19.5 = 3 + x$
    $x = \log 19.5 - 3$
    $x \approx -1.71$

15. a. $\log S = 4 + 2.3x$
    $S = 10^{4 + 2.3x}$
    $S = 10^4 \cdot 10^{2.3x}$

    b. $\log S = 200 + 1.5x$
    $S = 10^{200 + 1.5x}$
    $S = 10200 \cdot 101.5^x$

    c. $\log S = 500 - 1.8x$
    $S = 10^{500 - 1.8x}$
    $S = 10500 \cdot 10 - 1.8x$

Lesson 11 Land and Its Limits

Homework

1. a. 266 million, the current population
   b. (1,833,320 km^2 arable land)/(0.00073 km^2/person) = 2.5 billion people
   c. (1,833,320 km^2)/(0.0025 km^2/person) = 733 million people

Lesson 12 Resources and Limits

Homework

3. Doubling the reserves of a commodity will double the static index, because the static index is essentially a linear model of consumption, where each new year has the same consumption as the current year.

   Doubling the reserves of a commodity will not double the industrial index, because it is essentially an exponential model of consumption, with a fixed percent annual increase built in.

For example, double the reserves for Cobalt to 2,600,000 metric tons. The Static Index = 2,600,000/24,074 = 108 years. The Industrial Index = ln (0.06·108 + 1)/0.06 = 34 years. The Static Index doubled as expected. The Industrial Index increased by 10 years which is could have been approximated by ln 2/0.06 ≈ 11.6 years.

Lesson 13 The Sky is Falling

Homework

1. $OG = -0.0126x + 1.057$, where x is the number of years since 1940 and OG is the percent ore grade.

3. $OG = 1.053 \cdot (0.984)^x$, where x is the number of years since 1940 and OG is the percent ore grade.

8. Static index: 81 years. Only 81 years are added to the world supply of copper by mining a portion of the earth's crust. Industrial index: 41 years. If consumption increases by 3% per year, then only 41 years are added to the world supply of copper by mining a portion of the earth's crust.

Finance

Lesson 1 Payroll

Homework

1. a. Sue's withholding using the Percentage Method.
   $\$52.88 \cdot 2 = \$105.76$
   $\$1025 - \$105.76 = \$919.24$
   $\$118.35 + 0.28(919.24 - 913) = \$120.10$

3. Joe's Gross pay is \$675.
   His take home pay is $675 - 41.85 - 9.79 - 113 - 22.60 = \$487.76$.
   $487.76/675 = 0.7226$
   Joe takes home 72.26% of his gross pay.

5. a. Patrice will make about $\$7.50 \cdot 40 = \$300$ gross pay, resulting in $(0.28)(300) = \$216$ take home pay.
   b. Answers may vary, but Patrice will be working 40 hours per week for \$16, after taxes and day care are paid.
   d. What hourly wage would be needed for Patrice's take-home pay to be \$150 more than the cost of childcare?
   Let x be the hourly pay rate.
   Take-home pay is given by $40x - 0.28(40x) - 200 = 150$
   $$40x - 11.2x - 200 = 150$$
   $$28.8x - 200 = 150$$
   $$28.8x = 350$$
   $$x \approx \$12.15$$
   Graphical solution
   Graph $y_1 = 40x - 0.28(40x) - 200$
   $y_2 = 150$
   The two lines intersect at about (12.15, 150), showing that the hourly rate should be about \$12.15.

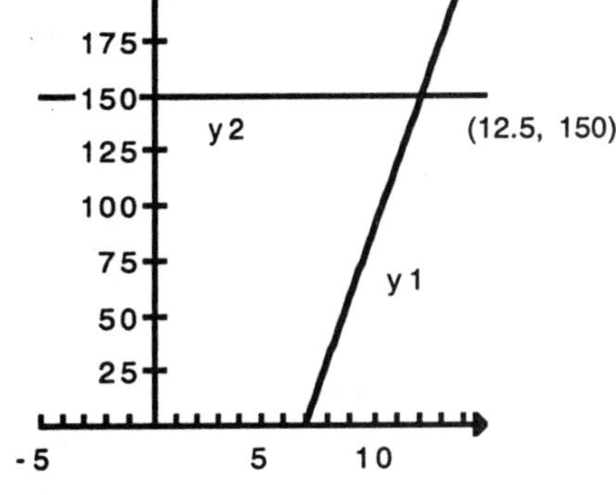

7. a. This year George gets a raise.
   \$42,000 4% pay raise makes $\$42,000 \cdot 0.04 = \$1680$ increase.
   $\$42,000 + 1680 = \$43,680$ gross pay for this year.
   b. George's 5% bonus is $43680(0.05) = \$2184$.
   But, he must pay about 32% of that bonus in salary deductions, leaving $2184 - (0.32)(2184) = \$1485.12$ to spend on Christmas gifts.
   c. Since George didn't receive his raise until July 1, his gross pay for the year would be \$21,000 for the first six months and \$21,000(1.04) for the second six months. $21,000 + (21,000)(1.04) =$ \$42,840 for the year.

His 5% bonus is 42,840 (0.05) = $2142.
But, he must pay about 32% of that bonus in salary deductions, leaving 2142 – (0.32)(2142) = $1,456.56 to spend on Christmas gifts.

Lesson 2 Simple Interest

Homework

1.

Account	PV	r	t	I
Joe	$675.78	4%	6 months	$13.51
Ted	$7,602.00	4%	3 months	$76.02
Sue	$842.10	3%	0.75 year	$18.95
Mary	$2,718.28	2.2%	12 months	$59.80

Joe: $I = PV \cdot r \cdot t = 675.78 \cdot 0.04 \cdot 0.05 = \$13.51$

Ted: $I = PV \cdot r \cdot t$
$76.02 = PV \cdot 0.04 \cdot 0.25$
$76.02 = 0.01 \cdot PV$
$\$7602 = PV$

Sue: $I = PV \cdot r \cdot t$
$18.95 = 842.10 \cdot 0.03 \cdot t$
$18.95 = 25.263 \cdot t$
$0.75 = t$, which is 9 months.

Mary: $I = PV \cdot r \cdot t$
$59.80 = 2718.28 \cdot r \cdot 1$
$0.022 = r$
Interest rate is 2.2%

3. $I = PV \cdot r \cdot t$
$39 = PV \cdot 0.052 \cdot 1$
$39 = 0.052PV$
$\$750 = PV$
Her principal is $750.

5. a. Interest earned on $5000 at 2% for 1 year:
$I = PV \cdot r \cdot t$
$I = 5000 \cdot 0.02 \cdot 1 = \$100$.
b. If you double the interest rate, you double the interest earned, so $200 is earned.

7.

Interest Rate	r	Interest Earned	Future Value
0%	0.00	$0	$1000.00
2%	0.02	$6.66	$1006.66
4%	0.04	$13.33	$1013.33
6%	0.06	$20.00	$1020.00
8%	0.08	$26.66	$1026.66
10%	0.10	$33.33	$1033.33

Keep in mind that banks truncate, they do not round.

Lesson 3 Annual Compound Interest

Nitty Gritty: Using Logs and Roots to Solve Equations

Practice:

1. Solve for t: $3 = 1.06^t$

$$\log 3 = \log 1.06^t$$
$$\log 3 = t \cdot \log 1.06$$
$$\frac{\log 3}{\log 1.06} = t$$
$$t \approx 18.85$$

3. Solve for x:

$$5000 = 2000 \cdot 1.043^x$$
$$5000 = 2000 \cdot 1.043^x$$
$$\frac{5000}{2000} = 1.043^x$$
$$\log 2.5 = \log 1.043^x$$
$$\log 2.5 = x \cdot \log 1.043$$
$$\frac{\log 2.5}{\log 1.043} = x$$
$$x \approx 21.76$$

5. Solve for t:

$$10{,}000 = 2000\left(1 + \frac{0.06}{12}\right)^{12t}$$
$$\frac{10{,}000}{2000} = \left(1 + \frac{0.06}{12}\right)^{12t}$$
$$\log 5 = \log\left(1 + \frac{0.06}{12}\right)^{12t}$$
$$\log 5 = 12t \cdot \log\left(1 + \frac{0.06}{12}\right)$$
$$\frac{\log 5}{\log\left(1 + \frac{0.06}{12}\right)} = 12t$$
$$\frac{\frac{\log 5}{\log\left(1 + \frac{0.06}{12}\right)}}{12} = t$$
$$t \approx 26.89$$

7. Solve for r: $1.3 = (1+r)^2$

$$\sqrt{1.3} = \sqrt{(1+r)^2}$$
$$\sqrt{1.3} = 1+r$$
$$\sqrt{1.3} - 1 = r$$
$$r \approx 0.14$$

9. Solve for x:

$200 = 100(1+x)^4$

$\sqrt[4]{2} = \sqrt[4]{(1+x)^4}$

$\sqrt[4]{2} = 1+x$

$\sqrt[4]{2} - 1 = x$

$x \approx 0.189$

11. Solve for r:

$1750 = 1000(1+\frac{r}{4})^{20}$

$\frac{1750}{1000} = (1+\frac{r}{4})^{20}$

$\sqrt[20]{1.75} = \sqrt[20]{(1+\frac{r}{4})^{20}}$

$\sqrt[20]{1.75} = 1+\frac{r}{4}$

$\sqrt[20]{1.75} - 1 = \frac{r}{4}$

$4\left(\sqrt[20]{1.75} - 1\right) = r$

$r \approx 0.114$

Homework

1.

Account	PV	r	t	FV
Joe	\$675.78	4%	6 years	\$855.07
Ted	\$7,479.71	4%	3 years	\$8,413.66
Sue	\$1,250	3%	5 years	\$1,450
Mary	\$2,700	2.5%	2 years	\$2,836.68

Joe: $FV = PV(1+r)^t$
$FV = 675.78(1+0.04)^6$
$FV = \$855.07$

Ted: $FV = PV(1+r)^t$
$8413.66 = PV(1+0.04)^3$
$8413.66 = PV(1.125)$
$\frac{8413.66}{1.125} = PV$
$\$7479.71 = PV$

Sue: $FV = PV(1+r)^t$
$1450 = 1250(1+0.03)^t$
$1.16 = (1.03)^t$
$\log 1.16 = \log(1.03)^t$
$\log 1.16 = t \cdot \log(1.03)$
$\frac{\log 1.16}{\log 1.03} = t$
$t \approx 5.02$ years

Mary: $FV = PV(1+r)^t$
$2836.68 = 2700(1+r)^2$
$\frac{2836.68}{2700} = (1+r)^2$
$\sqrt{\frac{2836.68}{2700}} = \sqrt{(1+r)^2}$
$\sqrt{\frac{2836.68}{2700}} = 1+r$
$\sqrt{\frac{2836.68}{2700}} - 1 = r$
$r \approx 0.025$ or 2.5%

3. Mary's daughter will get \$4051.63 on her 18th birthday.

$$FV = PV(1+r)^t$$
$$FV = 2000(1+0.04)^{18}$$
$$FV = 2000(1.04)^{18}$$
$$FV = \$4051.63$$

5. \$5,000 becomes \$10,000 in 5 years at about 14.9% interest.

$$FV = PV(1+r)^t$$
$$10000 = 5000(1+r)^5$$
$$2 = (1+r)^5$$
$$\sqrt[5]{2} = \sqrt[5]{(1+r)^5}$$
$$\sqrt[5]{2} = 1+r$$
$$\sqrt[5]{2} - 1 = r$$

$0.149 \approx r$, or about 14.9% interest.

7. a. It will take 72/8 = 9 years to double.
   b. It will take 72/8 = 9 years to double.
   c. It will take 72/8 = 9 years to double.
   d. The amount deposited (PV) has no effect on the doubling time; only the interest rate does.

11. a. $\frac{FV}{PV} = (1+r)^t$

$$\log\left(\frac{FV}{PV}\right) = \log(1+r)^t$$
$$\log\left(\frac{FV}{PV}\right) = t \cdot \log(1+r)$$
$$\frac{\log\left(\frac{FV}{PV}\right)}{\log(1+r)} = t$$

b.

$$t = \frac{\log\left(\frac{7500}{5000}\right)}{\log(1.032)} \approx 12.87 \text{ years.}$$

Lesson 4 General Compound Interest

Homework

1. $FV = PV\left(1+\frac{r}{m}\right)^{m \cdot t}$

$$FV = 14{,}500\left(1+\frac{0.035}{12}\right)^{12 \cdot 8}$$

$FV = \$19{,}177.56$ is the value of the account after 8 years.

3.

Compounded	Quarterly	Daily	Every second	Continuously
Future Value	$4,244.33	$4,257.05	$4,257.20	$4,257.20

Compounded Quarterly

$$FV = PV\left(1+\frac{r}{m}\right)^{m\cdot t}$$

$$FV = 3000\left(1+\frac{0.07}{4}\right)^{4\cdot 5}$$

$FV = \$4244.33$

Compounded Daily

$$FV = PV\left(1+\frac{r}{m}\right)^{m\cdot t}$$

$$FV = 3000\left(1+\frac{0.07}{365}\right)^{365\cdot 5}$$

$FV = \$4257.05$

Compounded Every second

$$FV = PV\left(1+\frac{r}{m}\right)^{m\cdot t}$$

$$FV = 3000\left(1+\frac{0.07}{31{,}536{,}000}\right)^{31{,}536{,}000\cdot 5}$$

$FV = \$4257.21$

Compounded Continuously

$$FV = PV\cdot e^{r\cdot t}$$

$$FV = 3000\cdot e^{0.07\cdot 5}$$

$$FV = PV\cdot e^{r\cdot t}$$
$$FV = 3000\cdot e^{0.07\cdot 5}$$
$$FV = \$4257.20$$

5. a. By the Rule of 72, Alisa's money would double in an estimated 72/9 = 8 years.

$$FV = PV\left(1+\frac{r}{m}\right)^{m\cdot t}$$

$$7110 = 3555\left(1+\frac{0.09}{12}\right)^{12\cdot t}$$

$$\frac{7110}{3555} = \left(1+\frac{0.09}{12}\right)^{12\cdot t}$$

$$\log 2 = \log(1.0075)^{12\cdot t}$$

$$\log 2 = 12\cdot t\cdot \log(1.0075)$$

$$\frac{\log 2}{12\cdot \log(1.0075)} = t$$

$t \approx 7.73$ or about 7 years 9 months.

b. How long will it take for her money to quadruple?

$$FV = PV\left(1+\frac{r}{m}\right)^{m\cdot t}$$

$$14220 = 3555\left(1+\frac{0.09}{12}\right)^{12\cdot t}$$

$$\frac{14220}{3555} = (1.0075)^{12\cdot t}$$

$$4 = (1.0075)^{12\cdot t}$$

$$\log 4 = \log(1.0075)^{12\cdot t}$$

$$\log 4 = 12t\cdot\log(1.0075)$$

$$\frac{\log 4}{12\cdot\log(1.0075)} = t$$

$t \approx 15.46$, or about 15 years 6 months.

7. a. Solve for PV: $FV = PV\left(1+\frac{r}{m}\right)^{m\cdot t}$

$$PV = \frac{FV}{\left(1+\frac{r}{m}\right)^{m\cdot t}}$$

or $PV = FV\left(1+\frac{r}{m}\right)^{-m\cdot t}$

8. a. $FV = PV\left(1+\frac{r}{m}\right)^{m\cdot t}$

$$\frac{FV}{PV} = \left(1+\frac{r}{m}\right)^{m\cdot t}$$

$$\log\left(\frac{FV}{PV}\right) = \log\left(1+\frac{r}{m}\right)^{m\cdot t}$$

$$\log\left(\frac{FV}{PV}\right) = mt\cdot\log\left(1+\frac{r}{m}\right)$$

$$\frac{\log\left(\frac{FV}{PV}\right)}{m\cdot\log\left(1+\frac{r}{m}\right)} = t$$

Lesson 5 Stream of Deposits

Homework

1. a 3, 15, 75, 375, … — It is geometric, with $r = 5$
   b 5, 9, 13, 17, … — It is not geometric
   c. 2.3, 3.91, 6.647, 11.2999, … — It is geometric, with $r = 1.7$

2. a. $a = 5$, $r = 2$ — 5, 10, 20, 40, 80
   b. $a = 20$, $r = 0.5$ — 20, 10, 5, 2.5, 1.25
   c. $a = \frac{3}{2}$, $r = -\frac{1}{2}$ — $\frac{3}{2}, \frac{-3}{4}, \frac{3}{8}, \frac{-3}{16}, \frac{3}{32}$

3. a. The sum of the first 10 terms of the sequence in problem 1(c).
$a = 2.3,\ r = 1.7,\ n = 10$

$$S_n = \frac{a(r^n - 1)}{r - 1}$$

$$S_{10} = \frac{2.3(1.7^{10} - 1)}{1.7 - 1} \approx 659.1122816$$

b. The sum of the first 15 terms of the sequence in problem 1(d).
$a = 64,\ r = 1/2,\ n = 15$

$$S_n = \frac{a(r^n - 1)}{r - 1}$$

$$S_{15} = \frac{64\left(\frac{1}{2}^{15} - 1\right)}{\frac{1}{2} - 1} \approx 127.9960938$$

5.

Deposit Amount	Interest Rate	Number of Periods per Year	Time	Future Value
\$25	9%	12	25 years	\$28,028.04
\$77.28	5%	12	10 years	\$12,000
\$750	7.2%	12	15 years	\$241,899.01
\$300.17	6.8%	4	12 years	\$22,000

Row 1: $$FV = R\frac{\left(1+\frac{r}{m}\right)^{mt} - 1}{\frac{r}{m}}$$

$$FV = 25\frac{\left(1+\frac{0.09}{12}\right)^{12\cdot 25} - 1}{\frac{0.09}{12}}$$

$FV = \$28,028.04$

Row 2: $$FV = R\frac{\left(1+\frac{r}{m}\right)^{mt} - 1}{\frac{r}{m}}$$

$$12,000 = R\frac{\left(1+\frac{0.05}{12}\right)^{12\cdot 10} - 1}{\frac{0.05}{12}}$$

$12,000 = R(155.28)$
$\$77.28 = R$

Row 3: $$FV = R\frac{\left(1+\frac{r}{m}\right)^{mt} - 1}{\frac{r}{m}}$$

$$FV = 750\frac{\left(1+\frac{0.072}{12}\right)^{12\cdot 15} - 1}{\frac{0.072}{12}}$$

$FV = \$241,899.01$

Row 4: $$FV = R\frac{\left(1+\frac{r}{m}\right)^{mt} - 1}{\frac{r}{m}}$$

$$22,000 = R\frac{\left(1+\frac{0.068}{4}\right)^{4\cdot 12} - 1}{\frac{0.068}{4}}$$

$22,000 = R(73.29)$
$\$300.17 = R$

7. a. The maximum monthly deposit is $\frac{\$2000}{12} \approx \$166.66$ per month.

b. At the end of 30 years, the account is worth:

$$FV = R\frac{\left(1+\frac{r}{m}\right)^{mt} - 1}{\frac{r}{m}}$$

$$FV = 166.66\frac{\left(1+\frac{0.05}{12}\right)^{12\cdot 30} - 1}{\frac{0.05}{12}} = \$138,704.22$$

c. If he makes the maximum monthly for 41 years, the account is worth:

$$FV = R\frac{\left(1+\frac{r}{m}\right)^{mt} - 1}{\frac{r}{m}}$$

$$FV = 166.66\frac{\left(1+\frac{0.05}{12}\right)^{12\cdot 41} - 1}{\frac{0.05}{12}} = \$269,384.74$$

d. The monthly deposit strategy is better than the annual one.

9. a.

Annual Interest Rate	Future Value
2%	$100\frac{\left(1+\frac{0.02}{23}\right)^{12\cdot 45} - 1}{\frac{0.02}{12}} = \$87,465.66$
3%	$100\frac{\left(1+\frac{0.03}{23}\right)^{12\cdot 45} - 1}{\frac{0.03}{12}} = \$114,037.29$
4%	$100\frac{\left(1+\frac{0.04}{23}\right)^{12\cdot 45} - 1}{\frac{0.04}{12}} = \$150,946.97$
5%	$100\frac{\left(1+\frac{0.05}{23}\right)^{12\cdot 45} - 1}{\frac{0.05}{12}} = \$202,643.72$

6%	$100\frac{\left(1+\frac{0.06}{23}\right)^{12\cdot 45}-1}{\frac{0.06}{12}} = \$275{,}599.26$
7%	$100\frac{\left(1+\frac{0.07}{23}\right)^{12\cdot 45}-1}{\frac{0.07}{12}} = \$379{,}259.46$
8%	$100\frac{\left(1+\frac{0.08}{23}\right)^{12\cdot 45}-1}{\frac{0.08}{12}} = \$527{,}453.98$
9%	$100\frac{\left(1+\frac{0.09}{23}\right)^{12\cdot 45}-1}{\frac{0.09}{12}} = \$740{,}487.84$
10%	$100\frac{\left(1+\frac{0.10}{23}\right)^{12\cdot 45}-1}{\frac{0.10}{12}} = \$1{,}048{,}250.17$
11%	$100\frac{\left(1+\frac{0.11}{23}\right)^{12\cdot 45}-1}{\frac{0.11}{12}} = \$1{,}494{,}841.32$

10. a. $$FV = R\frac{\left(1+\frac{r}{m}\right)^{mt}-1}{\frac{r}{m}}$$

$$FV\cdot\frac{r}{m} = R\left(\left(1+\frac{r}{m}\right)^{mt}-1\right)$$

$$\frac{FV\cdot\frac{r}{m}}{\left(\left(1+\frac{r}{m}\right)^{mt}-1\right)} = R$$

$$\frac{FV\cdot r}{m\left(\left(1+\frac{r}{m}\right)^{mt}-1\right)} = R$$

Lesson 6 Amortization

Homework

1.

Pres Value	Interest Rate	Loan Length	Payment	Total Amnt Paid	Finance Charge
$12,500	5.5%	60 months	$238.76	$14,325.60	$1,825.60
$118,700	6.25%	360 months	$730.86	$263,109.60	$144,409.60
$625.86	15%	30 months	$25.00	$750	$124.14
$9,800	4.85%	48 months	$225	$10,800	$1,000
$80,357.49	6.5%	180 months	$700	$126,000	$45,642.51

Row 1 Payment $R = PV\dfrac{\frac{r}{m}}{1-\left(1+\frac{r}{m}\right)^{-mt}}$

$$R = 12500\dfrac{\frac{0.055}{12}}{1-\left(1+\frac{0.055}{12}\right)^{-12\cdot 5}} = \$238.76$$

Total Amount Paid: $238.76·60 = $14,325.60
Finance Charge: $14,325.60 – 12,500 = $1,825.60

Row 2: Payment $R = PV\dfrac{\frac{r}{m}}{1-\left(1+\frac{r}{m}\right)^{-mt}}$

$$R = 118{,}700\dfrac{\frac{0.0625}{12}}{1-\left(1+\frac{0.0625}{12}\right)^{-12\cdot 30}} = \$730.86$$

Total Amount Paid: $730.86·360 = $263,109.60
Finance Charge: $263,109.60– 118,700 = $144,409.60

Row 3: Find t. You can use the guess-and-check method described in Activity 1 or you can graph $y_1 = 25$
and y_2 = 625.86(0.15/12)/(1–(1+0.15/12)^(–12·X))
and look for the x-value of the intersection point for the time in years.

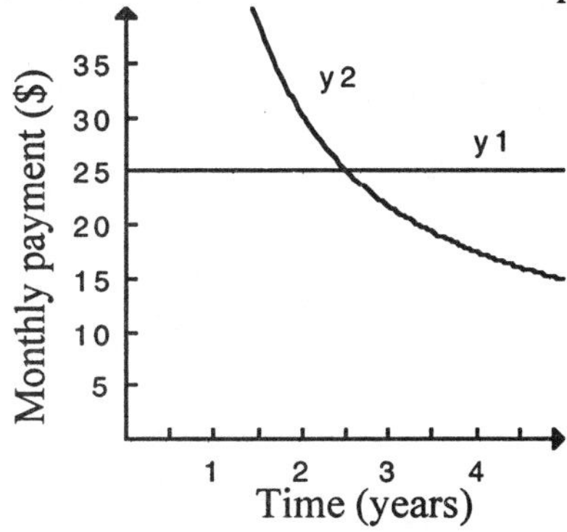

You can also use algebraic methods. The algebraic solution follows:

$$25 = 625.86\frac{\frac{0.15}{12}}{1-\left(1+\frac{0.15}{12}\right)^{-12\cdot t}}$$

$$25 = \frac{7.82325}{1-\left(1.0125\right)^{-12\cdot t}}$$

$$1-\left(1.0125\right)^{-12\cdot t} = \frac{7.82325}{25}$$

$$1-\left(1.0125\right)^{-12\cdot t} = 0.31293$$

$$\left(1.0125\right)^{-12\cdot t} = 0.68707$$

$$\log\left(1.0125\right)^{-12\cdot t} = \log 0.68707$$

$$-12\cdot t\cdot \log\left(1.0125\right) = \log 0.68707$$

$$t = \frac{\log 0.68707}{-12\cdot\log\left(1.0125\right)} \approx 2.52, \text{ or about 30 months.}$$

Total payments: 30 months · \$25 per month = \$750.00.
Finance charge: \$750 – \$625.86 = \$124.14

Row 4: Find r. You can use the guess-and-check method described in Activity 1 or
You can graph $y_1 = 225$
and y_2 = 9800(X/12)/(1-(1+X/12)^(-12·4))
Look for the intersection point. The x-value of the intersection point will be the interest rate.
You may want to zoom in on this intersection point to improve the accuracy.

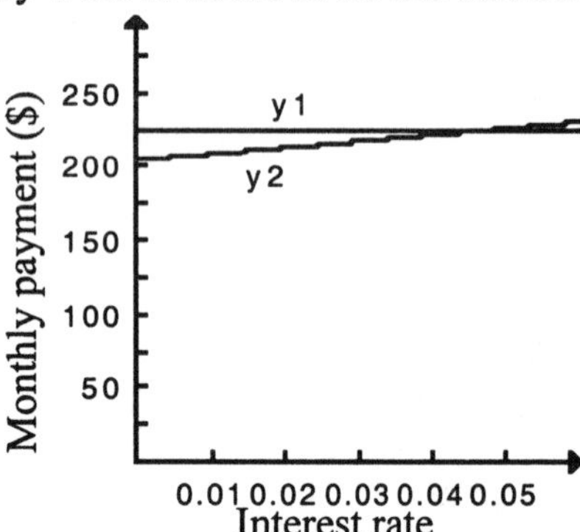

$r \approx 4.85\%$
Total amount paid: 225·48 = \$10800
Finance charge: \$10800 – \$9800 = \$1000

Row 5: Find PV. $R = PV\dfrac{\frac{r}{m}}{1-\left(1+\frac{r}{m}\right)^{-mt}}$

$$700 = PV\frac{\frac{0.065}{12}}{1-\left(1+\frac{0.065}{12}\right)^{-12\cdot 15}}$$

$$700\left(\frac{1-\left(1+\frac{0.065}{12}\right)^{-12\cdot 15}}{\frac{0.065}{12}}\right) = PV$$

$\$80357.49 = PV$
Total amount paid: $700\cdot 180 = \$126{,}000$
Finance charge: $\$126{,}000 - \$80{,}357.49 = \$45{,}642.51$

3. a. The price of the house less the down payment is the mortgage amount. $185,000 – $22,500 = $162,500, the amount they are financing.
   b. The monthly payment for a 30 year loan is:

$$R = PV\frac{\frac{r}{m}}{1-\left(1+\frac{r}{m}\right)^{-mt}}$$

$$R = 162{,}500\frac{\frac{0.0675}{12}}{1-\left(1+\frac{0.0675}{12}\right)^{-12\cdot 30}} = \$1053.97 \text{ per month}$$

   c. The monthly payment for the 15-year loan is:

$$R = 162{,}500\frac{\frac{0.0625}{12}}{1-\left(1+\frac{0.0625}{12}\right)^{-12\cdot 15}} = \$1393.31 \text{ per month}$$

5. d. Payment 4

   First, calculate the interest. $I = \$590.44 \cdot 0.12 \cdot \dfrac{1}{12} = \$5.90$
   Second, find the amount of the $77.16 payment that is applied to principal, by subtracting the interest from that payment. $77.16 - $5.90 = $71.26.
   Third, find the remaining balance by subtracting the amount applied to principal from the previous remaining balance. $590.44 – $71.26 = $519.18

Payment 5

First, calculate the interest. $I = \$519.18 \cdot 0.12 \cdot \frac{1}{12} = \$5.19$

Second, find the amount of the \$77.16 payment that is applied to principal, by subtracting the interest from that payment. \$77.16 - \$5.19 = \$71.97.
Third, find the remaining balance by subtracting the amount applied to principal from the previous remaining balance. \$519.18 – \$71.97 = \$447.21

Payment 6

First, calculate the interest. $I = \$447.21 \cdot 0.12 \cdot \frac{1}{12} = \$4.47$

Second, find the amount of the \$77.16 payment that is applied to principal, by subtracting the interest from that payment. \$77.16 - \$4.47 = \$72.69.
Third, find the remaining balance by subtracting the amount applied to principal from the previous remaining balance. \$447.21 – \$72.69 = \$374.52

Payment 7

First, calculate the interest. $I = \$374.52 \cdot 0.12 \cdot \frac{1}{12} = \$3.75$

Second, find the amount of the \$77.16 payment that is applied to principal, by subtracting the interest from that payment. \$77.16 - \$3.75 = \$73.41.
Third, find the remaining balance by subtracting the amount applied to principal from the previous remaining balance. \$374.52 – \$73.41 = \$301.11

Payment 8

First, calculate the interest. $I = \$301.11 \cdot 0.12 \cdot \frac{1}{12} = \$3.01$

Second, find the amount of the \$77.16 payment that is applied to principal, by subtracting the interest from that payment. \$77.16 - \$3.01 = \$74.15.
Third, find the remaining balance by subtracting the amount applied to principal from the previous remaining balance. \$301.11 – \$74.15 = \$226.96

Payment 9

First, calculate the interest. $I = \$226.96 \cdot 0.12 \cdot \frac{1}{12} = \$2.27$

Second, find the amount of the \$77.16 payment that is applied to principal, by subtracting the interest from that payment. \$77.16 - \$2.27= \$74.89.
Third, find the remaining balance by subtracting the amount applied to principal from the previous remaining balance. \$226.96– \$74.89= \$152.07

Payment 10

First, calculate the interest. $I = \$152.07 \cdot 0.12 \cdot \frac{1}{12} = \$1.52$

Second, find the amount of the \$77.16 payment that is applied to principal, by subtracting the interest from that payment. \$77.16 - \$1.52= \$75.64.
Third, find the remaining balance by subtracting the amount applied to principal from the previous remaining balance. \$152.07– \$75.64= \$76.43

e. Last payment

First calculate the interest during the last month. $I = 76.43 \cdot 0.12 \cdot \frac{1}{12} = \$0.76$

Second, add the interest to the remaining principal. \$75.64 + \$0.76 = \$76.43.

Payment Number	Payment amount	Monthly Interest	Amount applied to principal	Balance remaining
--	--	--	--	$800.00
1	$77.16	$8.00	$69.16	$730.84
2	$77.16	$7.31	$69.85	$660.99
3	$77.16	$6.61	$70.55	$590.44
4	$77.16	$5.90	$71.26	$519.18
5	$77.16	$5.19	$71.97	$447.21
6	$77.16	$4.47	$72.69	$374.52
7	$77.16	$3.75	$73.41	$301.11
8	$77.16	$3.01	$74.15	$226.96
9	$77.16	$2.27	$74.89	$152.07
10	$77.16	$1.52	$75.64	$76.43
11	$77.19	$0.76	$76.43	$00.00

Lesson 7 Retirement

Homework

1. Use this formula to find the missing values in the table. $R = PV\dfrac{\frac{r}{m}}{1-\left(1+\frac{r}{m}\right)^{-mt}}$

Row 1: $R = 880{,}755\dfrac{\frac{0.055}{12}}{1-\left(1+\frac{0.055}{12}\right)^{-12\cdot 20}}$ = \$6058.81 per month

Row 2: $R = 425{,}323\dfrac{\frac{0.0625}{12}}{1-\left(1+\frac{0.0625}{12}\right)^{-12\cdot 15}}$ = \$3646.81 per month

Row 3: $7850 = 625{,}894\dfrac{\frac{0.12}{12}}{1-\left(1+\frac{0.12}{12}\right)^{-12\cdot t}}$

Solve by graphing $y_1 = 7859$
y_2 = 625,894(0.12/12)/(1-(1+0.12/12)^(-12x))
Find the intersection point: (13.367337, 7859)

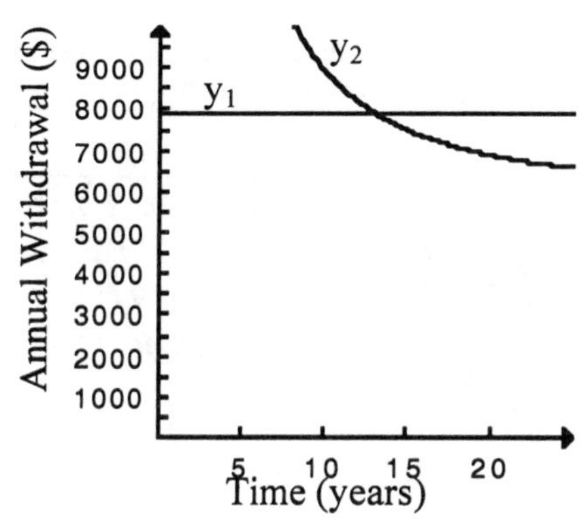

Or use algebra to solve for t.

$$\frac{7850}{625{,}894} = \frac{0.01}{1-(1.01)^{-12\cdot t}}$$ multiply by 625,894; simplify

$$7850\cdot\left(1-(1.01)^{-12\cdot t}\right) = 0.01\cdot 625{,}894$$ Cross multiply

$$7850 - 7850(1.01)^{-12\cdot t} = 6258.94$$ Distribute and simplify

$$-7850(1.01)^{-12\cdot t} = -1591.06$$ Subtract 7850

$$(1.01)^{-12\cdot t} = 0.2026828$$ Divide by –7850

$$\log(1.01)^{-12\cdot t} = \log(0.2026828)$$ Take log of both sides

$$-12t\cdot\log(1.01) = \log(0.2026828)$$ Property of logs

$$t = \frac{\log(0.2026828)}{-12\cdot\log(1.01)}$$ Divide by –12·log(0.2026828)

$t \approx 13.36$ or 13 years 4 months

Row 4: $$3500 = 285{,}653\frac{\frac{r}{12}}{1-\left(1+\frac{r}{12}\right)^{-12\cdot 10}}$$

Solve by graphing $y1 = 3500$
$y2 = 285653(X/12)/(1-(1+X/12)^{\wedge}(-12\cdot 20))$
Or simplify with some algebra first.

$$\frac{3500}{285{,}653} = \frac{\frac{r}{12}}{1-\left(1+\frac{r}{12}\right)^{-12\cdot 10}}$$ Divide by 285653

$$3500 - 3500\left(1+\frac{r}{12}\right)^{-120} = 23804.42r$$ Cross multiply

$$-3500\left(1+\frac{r}{12}\right)^{-120} = 23804.42r - 3500$$ Subtract 3500

Now, graph the left side of the equation as y_1 and the right side of the equation as y_2. Look for the intersection point, (0.08226228, 0.01225263). The rate is about 8.226%.

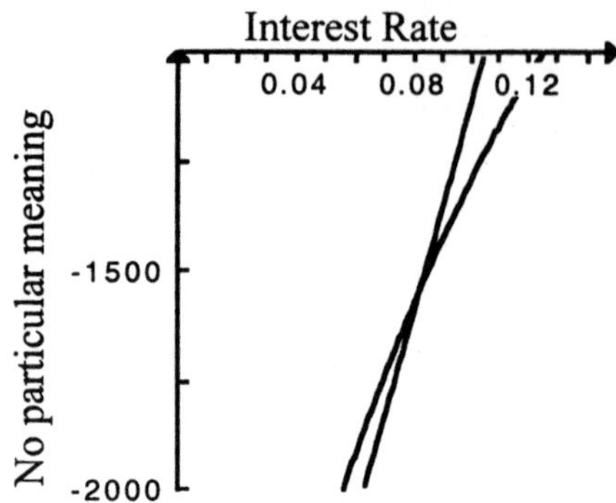

Row 5: $7000 = PV\dfrac{\dfrac{0.065}{12}}{1-\left(1+\dfrac{0.065}{12}\right)^{-12\cdot 25}}$

$$7000\cdot\frac{1-\left(1+\dfrac{0.065}{12}\right)^{-12\cdot 25}}{\dfrac{0.065}{12}} = PV$$ multiply by the reciprocal

$\$1{,}036{,}718.86 = PV$

Present Value	Interest Rate	Time (in years)	Withdrawal
$880,755.00	5.5%	20	$6,058.60
$425,323.00	6.25%	15	$3,646.81
$625,894.00	12%	13 yr. 4 mo.	$7,850.00
$285,653.00	8.23%	10	$3,500.00
$1,036,718.86	6.5%	25	$7,000.00

3. This is a simple interest problem. We just need to calculate the interest that $924,000 will earn in one month.

   $I = 924{,}000\cdot 0.08\cdot\dfrac{1}{12} = \$6{,}160.$

4. This is a simple interest problem. We just need to calculate the principal that will generate $8400 per month in interest.

   $8400 = PV\cdot 0.07\cdot\dfrac{1}{12}$

   $\dfrac{8400\cdot 12}{0.07} = PV$

   $\$1{,}440{,}000 = PV$, the principal that Lou will need.

5. a. Use the following formula to calculate the monthly lease payment on a car.

$$FV = PV\left(1+\frac{r}{m}\right)^n - R\frac{\left(1+\dfrac{r}{m}\right)^n - 1}{\dfrac{r}{m}}$$

$$12{,}000 = 20{,}000\left(1+\frac{0.035}{12}\right)^{36} - R\frac{\left(1+\dfrac{0.035}{12}\right)^{36} - 1}{\dfrac{0.035}{12}}$$

$$12{,}000 - 20{,}000\left(1+\frac{0.035}{12}\right)^{36} = -R\frac{\left(1+\dfrac{0.035}{12}\right)^{36} - 1}{\dfrac{0.035}{12}}$$

$-10210.81752 = -R\cdot 37.89973$

$\$269.42 = R$

This is a monthly lease payment of $269.42

b. Use the formula that relates regular payments (R), and present value (PV).

$$R = PV\frac{\frac{r}{m}}{1-\left(1+\frac{r}{m}\right)^{-mt}}$$

$$R = 20{,}000\frac{\frac{0.035}{12}}{1-\left(1+\frac{0.035}{12}\right)^{-12\cdot 5}}$$

$R =$ \$363.83 per month car payment.

6. a. You will get $\frac{1}{20}$ of the \$17,000,000 jackpot each year.

$\frac{1}{20}\cdot 17{,}000{,}000 = \$850{,}000$ per year.

b. If you select "lump sum" and win (not very likely), you need to calculate what you will receive. Calculate the present value of a stream of deposits where the 20 annual deposits are the payments found in part (a). Assume 6% compounded annually.
On the other hand, you know that you could get regular payments of \$850,000 per year. This question is asking for the present value of those payments. So, we need a formula that relates present value (PV) with regular payments (R).

$$R = PV\frac{\frac{r}{m}}{1-\left(1+\frac{r}{m}\right)^{-mt}}$$

$$850{,}000 = PV\frac{\frac{0.06}{1}}{1-\left(1+\frac{0.06}{1}\right)^{-1\cdot 20}}$$

$850{,}000 = PV\cdot 0.08718$

$\$9{,}749{,}433.03 = PV$

The lump sum payment is \$9,749,433.03

Functions

Lesson 1 What Is a Function?

Homework

1. K is not a function because two different ordered pairs have the same first member: (0,1) and (0,5).
3. M is a function because no two ordered pairs have the same first member.
5. O is a function because no two ordered pairs have the same first member.
7. This Time vs. Temperature table is a function because each different time in the first column has one and only one temperature in the second column.
10. This Super Bowl Vs. MVP table is not a function because there are two identically numbered Super Bowls in the first column that have different MVP's in the second column. XII -> Randy White and XII -> Harvey Martin.

Lesson 2 Four Representations of a Function

Homework

1. This is a function. No vertical line crosses the graph more than once.
3. Not a function. Vertical lines cross the graph more than once.

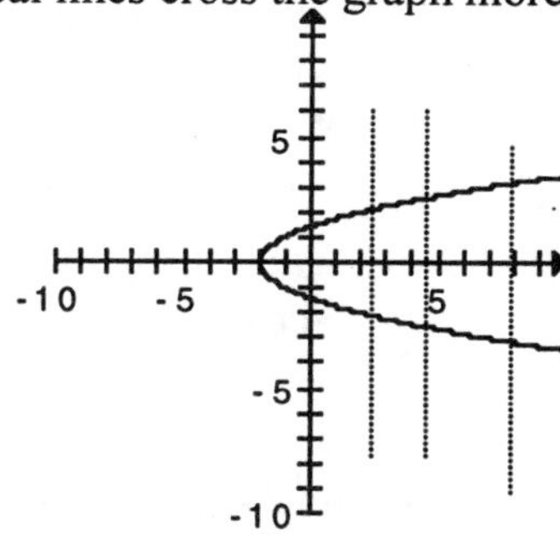

5. a. Integers b. Integers c. Domain is {–1, 2, 3, 5} d. Range is {3, 6, 7, 9}
   e. The process is to add 4 to each element in the domain to produce the elements in the range.
9. a. Capital letters b. Lower-case letters c. Domain is {A, B, C} d. Range is {a, b, c, d}
   e.

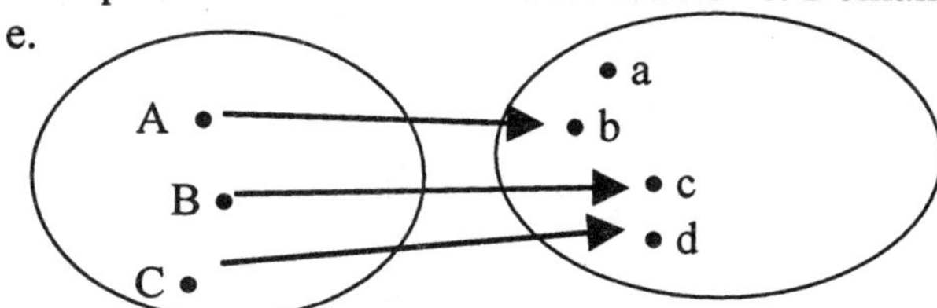

13. It is a function.

Numerical

x	y
–2	9
–1	6
0	5
1	6
2	9

No pair of y-values has the same x-value.

Graphical

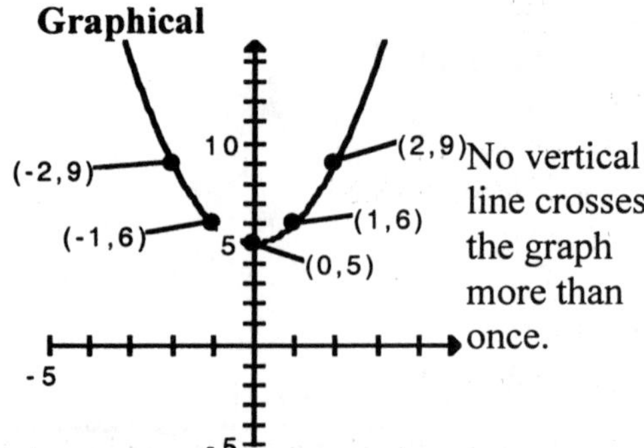

No vertical line crosses the graph more than once.

Symbolic

$y = x^2 + 5$

There is no ± on the right side, which means that any x-value produces only one y-value.

Verbal

The process asks us to square a number, then increase this result by 5. This gives a unique new number each time.

14. It is not a function.

Numerical

x	y
0	0
1	–1
1	1
4	–2
4	2

No pair of y-values has the same x-value.

Graphical

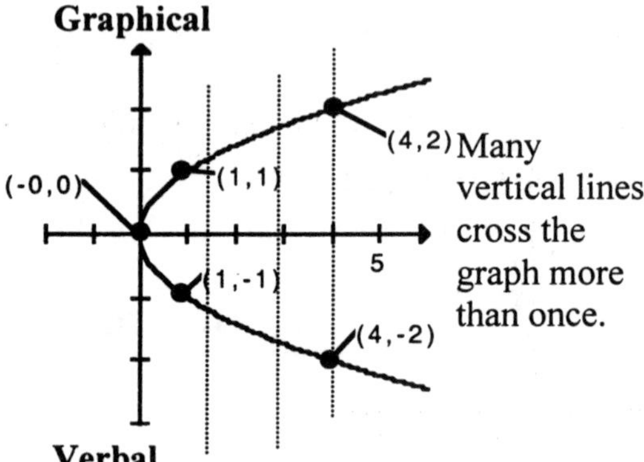

Many vertical lines cross the graph more than once.

Symbolic

$y = \pm\sqrt{x}$

A ± on the right side means that each x-value produces both a positive and negative number for y.

Verbal

The process asks us to produce the principal and negative root for all numbers greater than zero. Each input produces ambiguous output because it could be positive or negative.

16. A vertical line super-imposed on a graph traces out all possible ordered pairs that have the same first element. Functions require that only one of these ordered pairs be used. This means that any vertical line that crosses the graph more than once describes a relation, not a function.

Lesson 3 Function Notation

Homework

2. a. $h(0) = \dfrac{6-3(0)}{2} = 3$

b. $h(4) = \dfrac{6-3(4)}{2} = -3$

c. $h(-5) = \dfrac{6-3(-5)}{2} = 10.5$

d. What is w if $h(w) = -15$?

$$-15 = \frac{6-3w}{2}$$

$-30 = 6 - 3\text{w}$

$36 = -3\text{w}$ or $w = 12$

e. What is w if $h(w) = 13.5$?

$$13.5 = \frac{6-3w}{2}$$

$27 = 6 - 3w$

$21 = -3w$ or $w = -7$

5. $d(4)$ would represent the distance traveled after 4 hours.

7. a. $h(a) = \text{w}$ b. $h(b) = \text{k}$

c. Many answers are possible. For example, $x = \text{b}$ and $x = \text{c}$

d. When $h(x) = \text{t}$, $x = \text{d}$

Lesson 4 Families of Functions

Homework

1. $f(x) = 2x + 3$, an increasing linear function

$f(-6) = 2(-6)+3 = -12+3 = -9$

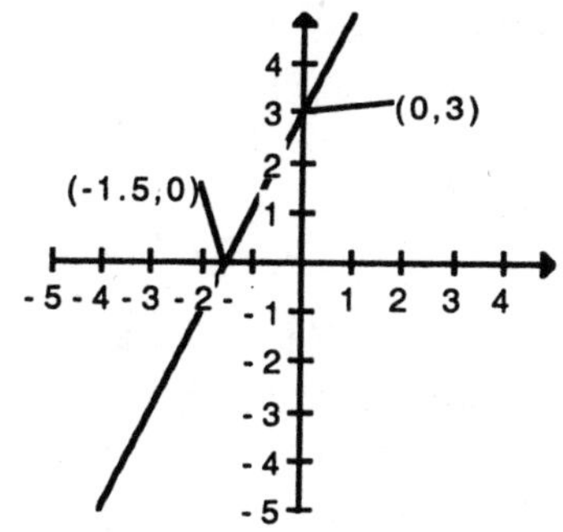

5. $f(x) = \sqrt{x} + 3$ Increasing square root function

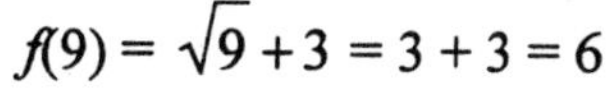

$f(9) = \sqrt{9} + 3 = 3 + 3 = 6$

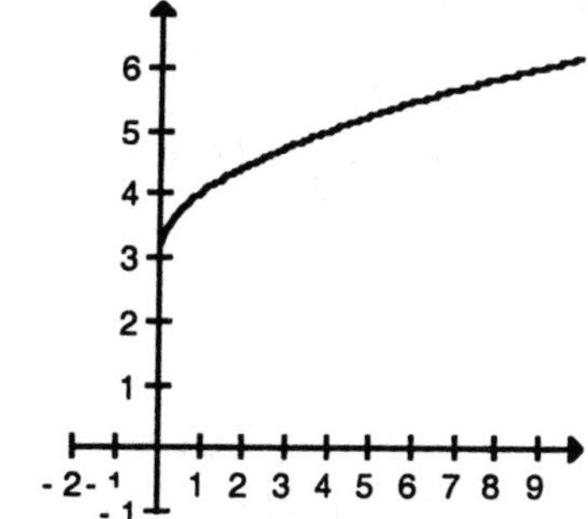

9. $f(x) = \ln(x+3)$, Increasing log function1

$f(0) = \ln(3) \approx 1.0986$

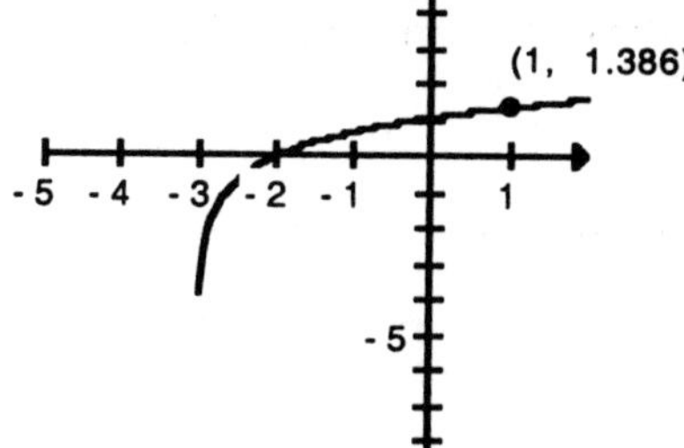

12. $f(x) = 10^{-x}$, Decreasing exponential function

$f(4) = 0.0001$

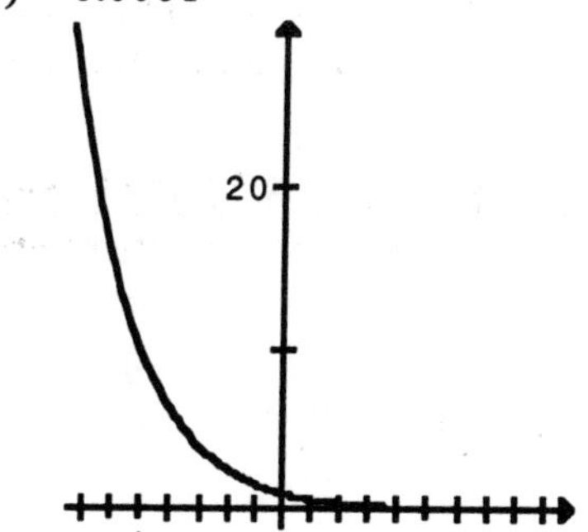

Lesson 5 Modeling Data with Functions

Homework

1. Using 1900 as $t = 0$, the linear regression model gives $P(t) = 43.88+2.29t$
   a. In our model $P(1950)$ is $P(50) = 158.38$ million people
   b. In our model $P(1985)$ is $P(85) = 238.53$ million people
   c. Using trace to approximate t when $P(t) = 200$ million shows that $t = 68$. There were 200 million people in 1968.
   d. $P(112) \approx 300$ million. The US would reach a population of 300 million in the year 2012.

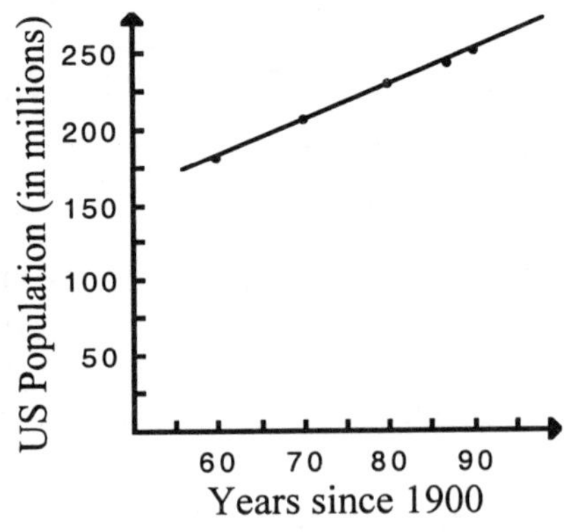

3. Using 1900 as $t = 0$, the exponential regression model gives: $R(t)$=155.242(1.0834^t)
   a. In our model $R(1955)$ is $R(55) = \$12,707.76$
   b. In our model $R(2010)$ is $R(110) = \$1,040,178.25$
   c. For $R(t) = 200000$, use trace to get $R(89) \approx 200,000$, so 1989 is the year.
   d. For $R(t) = \%500,00$, change Ymax to 500,000 and Xmax to 110. The trace gives $R(101) \approx \$500,000$, so 2001 is the year.

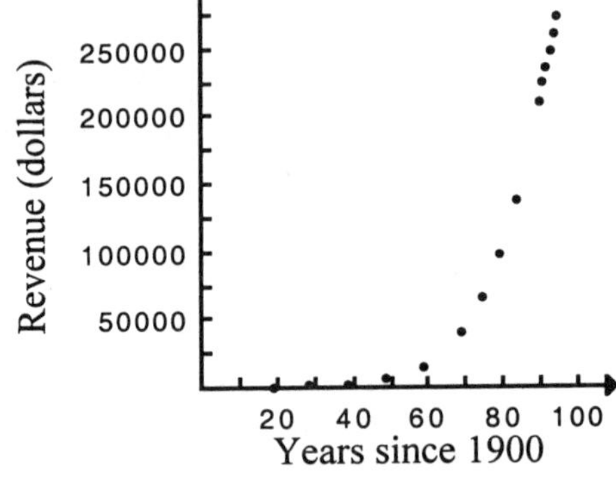

4. Using 1900 as $t = 0$, the cubic regression model is:
   $H(t) = 410.148t^3 - 103466.167t^2 + 8723132.684t - 242726754.1$
   a. In our model $H(1965)$ is $H(65) = -230788$
   b. This negative answer for the number of houses sold does not make sense, so this model fails to extrapolate reliably before 1970.
   c. $H(66) = 217,000$. In about 1966.
   d. For $H(t) = 4,500,000$, use trace. The result is that $H(98) \approx 4,500,000$. So 1998 was the year at 4,500,000 houses sold. The answers for d. depends on the use of the cubic model to fit the data. Other models chosen to fit the data will give different answers, especially when predicting beyond the data set.

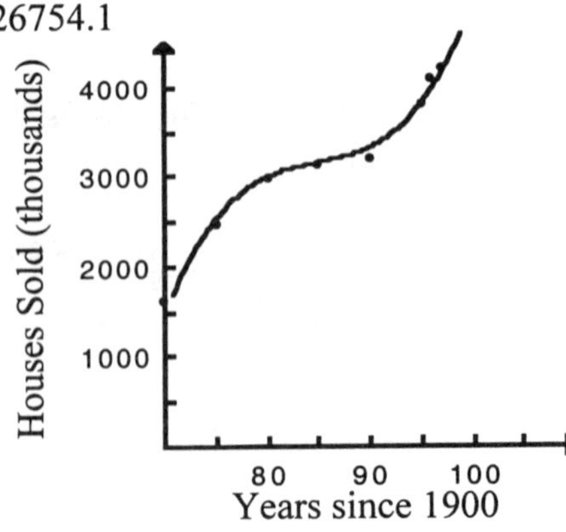

Lesson 6 Stopping Distance

Homework

5. a.

Water Temp. (C°)	Swimming Speed (cm/sec)	1st Difference	2nd difference
5	15		
10	27.5	$27.5 - 15 = 12.5$	
15	32	$32 - 27.5 = 4.5$	$4.5 - 12.5 = -8$
20	30.5	$30.5 - 32 = -1.5$	$-1.5 - 4.5 = -6$
25	22	$22 - 30.5 = -8.5$	$-8.5 - (-1.5) = -7$
30	$22 - 15.5 = 6.5$	$-7 + -8.5 = -15.5$	-7

Using the method of Activity 2 we can work backwards to find the swimming speed at 30° C if the second differences are constant at 7.

b. $y = -0.137x^2 + 4.454x - 3.7$

c. The shape of the quadratic function is a parabola which opens down. Fish swim slowly when water is cold, faster as the water temperature increases, but slow again when the temperature is too high.

d. The fish will swim fastest at the vertex of the parabola. Using trace gives $f(16.24°) \approx 32.45$ cm/sec. The fish swim fastest at a water temperature of 16.24° C.

e. Use the value feature of the graphing utility to find that f(30) = 6.5 cm/sec, which confirms the estimate in part a.

8.

I (amps)	2	4	5
P (watts)	176	224	200

a. $P(I) = -16I^2 + 120I$

b. Using the value command of the graphing utility gives:

$P(1) = 104$ $P(2) = 176$ $P(3) = 216$ $P(4) = 224$ $P(5) = 200$

A setting of 4 amps yields the most power at 224 watts.

Lesson 7: The Allegory of the Cave

Homework

3. For circular cone, with vertex pointing up.

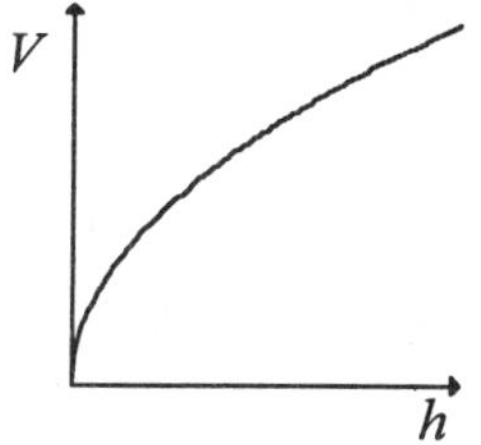

5. For the double cone,

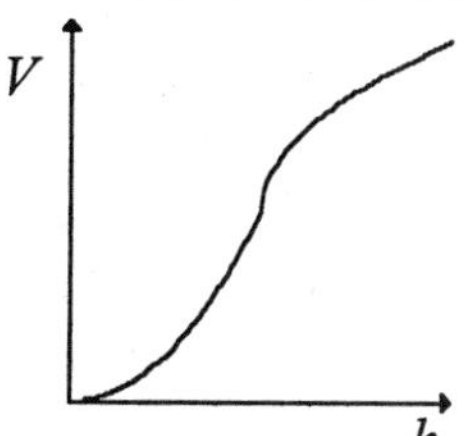

7. Wherever you create a function where y = a constant, unchanging value, all values for y *will* be the same no matter what the independent variable, x, is.

Lesson 8: Transformations of Functions

Homework

1. Assume $t = 0$ in 1900.
   b. Using linear regression, the function for the minimum wage data is $f(x) = 0.109x - 5.7$
2. b. The "new" minimum wage data is plotted above in 1.a. This data is shifted up from the original data by \$1.15.
   c. $f(x) = 0.109x - 4.55$. This function has the same slope as the original function, but has a different constant value. That value differs by exactly \$1.15.

4. d. $f(x - 2)$

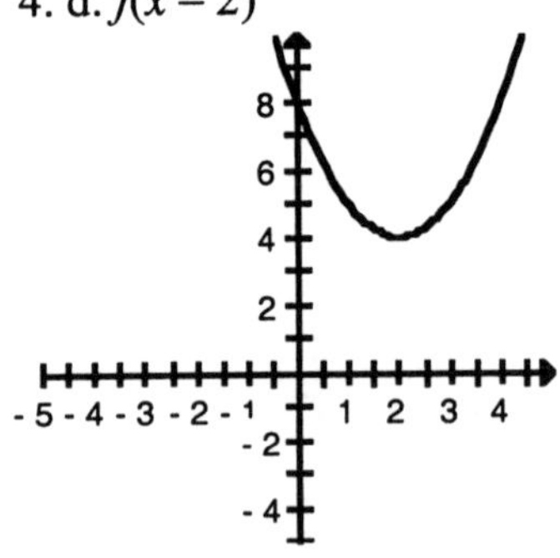

Lesson 9: More Transformations of Functions

Homework

2. a. $f(x) + 2$

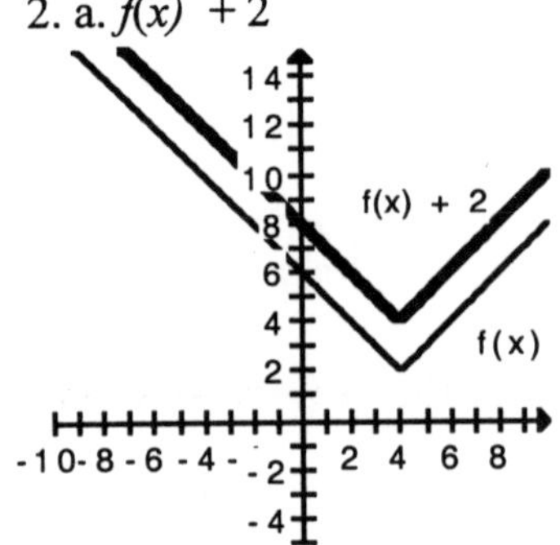

b. $f(x) - 4$

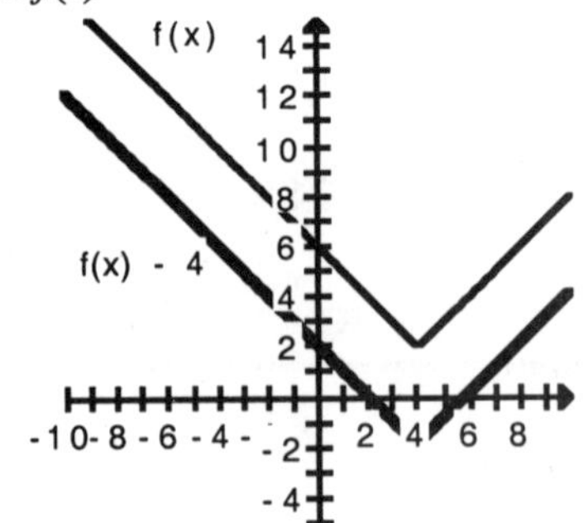

c. $f(x + 4)$

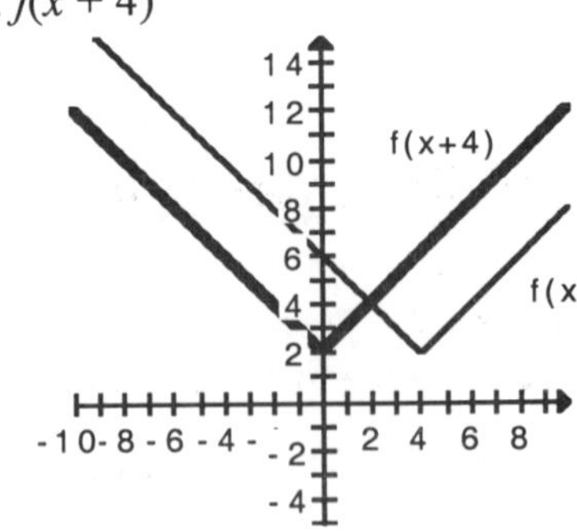

Lesson 10: Functions at the Scene of an Accident

Homework

1. Answers will vary. If the original relationship is a function and the inverse relationship is a function, then each element of the domain must correspond to one and only one element of the range.

5. The horizontal line test. If two x-values have the same y-value, an inverse function will not exist.

7. No horizontal line will cross the graph more than once. Therefore, an inverse function exists.

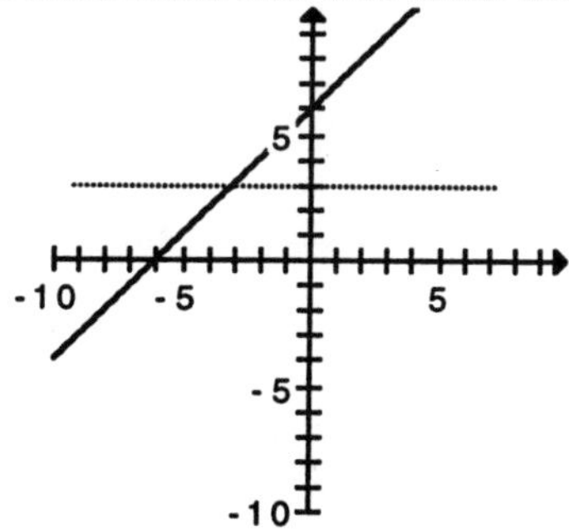

9. The inverse does not exist because a horizontal line crosses the graph more than once.

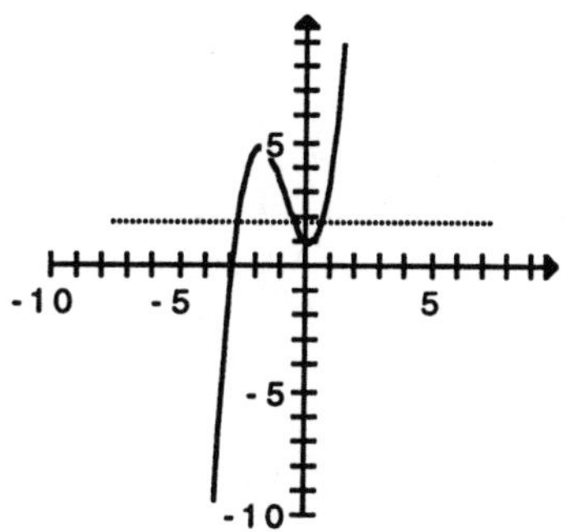

Nitty Gritty: Solving Equations

1. Solve for y:

$8x + 2y = -6$	
$2y = -6 - 8x$	Subtract $8x$ from both sides
$y = -3 - 4x$	Divide both sides by 2

3. Solve for y:

$x + 2y = 5$	
$2y = 5 - x$	Subtract x from both sides
$y = \frac{5}{2} - \frac{1}{2}x$	Divide both sides by 2

5. Solve for y:

$-3 = 3x - 3y$	
$3y = 3x + 3$	Add $3y$ *to* both sides
$y = x + 1$	Divide both sides by 3

7. Solve for m:

$\mathrm{e} = \mathrm{mc}^2$	
$\frac{e}{c^2} = \mathrm{m}$	Divide both sides by c^2

9. Solve for l:

$\mathrm{P} = 2l + 2w$	
$\mathrm{P} - 2w = 2l$	Subtract $2w$ from both sides
$\frac{P - 2w}{2} = l$	Divide both sides by 2

11. Solve for y:

$x^2 = y^2 + 7$	
$x^2 - 7 = y^2$	Subtract 7 from both sides
$y = \pm\sqrt{x^2 - 7}$	Take the principal and negative square root

13. Solve for c:

$e = mc^2$

$\frac{e}{m} = c^2$ Divide both sides by m

$\pm\sqrt{\frac{e}{m}} = c$ Take the principal and negative square root

15. Solve for r:

$A = \pi r^2$

$\frac{A}{\pi} = r^2$ Divide both sides by π

$\pm\sqrt{\frac{A}{\pi}} = r$ Take the principal and negative square root

17. Solve for *g*:

$H = \frac{1}{2}gt^2$

$2H = gt^2$ Multiply both sides by 2

$\frac{2H}{t^2} = g$ Divide both sides by t^2

19. Solve for *A*:

$Aa + b = 1$

$Aa = 1 - b$ Subtract *b* from both sides 2

$A = \frac{1-b}{a}$ Divide both sides by *a*

Lesson 11: Inverse Functions

Homework

2. $f(x) = \frac{2}{3}x - 7$

$y = \frac{2}{3}x - 7$ replace $f(x)$ with y

$x = \frac{2}{3}y - 7$ interchange x and y

$x + 7 = \frac{2}{3}y$ solve for y *by* first adding 7 to each side

$\frac{3}{2}x + \frac{21}{2} = y$ multiply both sides by 3/2

$f^{-1}(x) = \frac{3}{2}x + \frac{21}{2}$ replace y *with* $f^{-1}(x)$

3. $f(x) = 2x^2 + 7, x \geq 0$

$y = 2x^2 + 7$ replace $f(x)$ with y

$x = 2y^2 + 7$ interchange x and y

$x - 7 = 2y^2$ solve for y *by* first subtracting 7 from each side

$\frac{x-7}{2} = y^2$ divide both sides by 2

$-\sqrt{\frac{x-7}{2}} = y$ take the negative square root of left side to solve for y (because $x \leq 0$)

$f^{-1}(x) = -\sqrt{\frac{x-7}{2}}$ replace y *with* $f^{-1}(x)$

5. $f(x) = \sqrt{x} + 3, \ x \geq 0$

$y = \sqrt{x} + 3$	replace *f(x)* with y
$x = \sqrt{y} + 3$	interchange *x* and y
$x - 3 = \sqrt{y}$	solve for *y by* first subtracting 3 from each side
$(x-3)^2 = y$	square both sides
$f^{-1}(x) = (x-3)^2$	replace *y with* $f^{-1}(x)$

Geometry

Lesson 1 Where in the World Is Geometry?

Homework

3. This is false. Not all rectangles are squares.

4. This is true. All squares are rectangles.

Nitty Gritty: How to Use a Protractor

1. 52°
3. 46°
5. 136°
7. 120°
9. Each angle is 90°.

Lesson 2 How Do You Measure Up?

Lesson 3 Do You Have the Time?

Activity 2–Time Zones

4.

City		Day: Time	Approximate Angle of Rotation
A	Phoenix	Wednesday: noon	0°
B	Los Angeles	Wednesday: 11am	15°
C	Chicago	Wednesday: 1pm	15°
D	New York	Wednesday: 2pm	30°
E	Anchorage	Wednesday: 9am	45°
F	Lisbon	Wednesday: 7pm	105°
G	Jakarta	Thursday: 2am	210°
H	Sydney	Thursday: 5am	255°
I	Kuwait City	Wednesday: 10pm	150°
J	Beijing	Thursday: 3am	225°
K	Buenos Aires	Wednesday: 3pm	45°
L	Bogota	Wednesday: 2pm	30°
M	Capetown	Wednesday: 8pm	120°
N	Rome	Wednesday: 8pm	120°
O	Moscow	Wednesday: 9pm	135°
P	Cairo	Wednesday: 9pm	135°

Homework

2. a. 46° and 134° linear pair, supplementary angles
   b. 62°, *a* and *b* are vertical angles
   c. 66° and 24° complementary angles

3. If the sun sets at 6:30 pm in Phoenix, Arizona, the sun will set at approximately 6:40 pm in Yuma, Arizona. Yuma has an approximate longitude of 114°40'W. Phoenix has an approximate longitude of 112°10'W. This means that Yuma is 2.5° further west than Phoenix. Since the sun's apparent westward movement is 15° per hour, the sun will cross the western horizon in Yuma 2.5/15 of an hour later than in Phoenix, or about 10 minutes.

Lesson 4 How Do I Get There from Here?

Homework

3. a.

Compass Direction	Bearing
North	000
Northeast	045
East	090
Southeast	135
South	180
Southwest	225
West	270
Northwest	315

   b. Bearings are better than just the eight compass labels. Bearings are 3-digit numbers that give much better accuracy.

   c. 1. 045 2. 315 3. 315 4. 135

   d. Bearings and directions in orienteering both express a good numerical accuracy. Bearings would be better if a succinct numerical record is needed. Directions in orienteering are better for actually describing a sense of direction from a given point.

Lesson 5 I'm Going in Circles!

Homework

2.

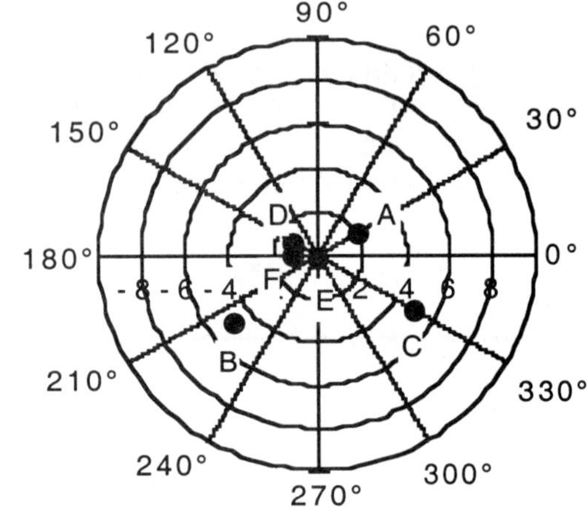

Lesson 6 Hot Times in the City

Homework

1. Make the angles 38°.

2. Make the first bounce at 32° and the second 58°.

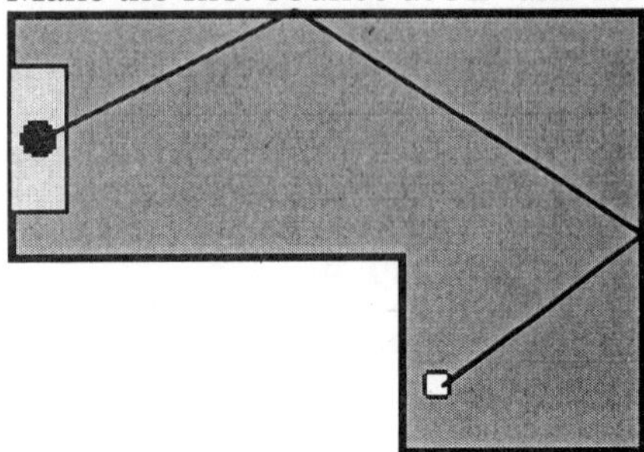

Lesson 7 The Shape of Things

Nitty Gritty: Solving Proportions

1. $\frac{12}{9} = \frac{x}{3}$
By inspection
$x = 4$

3. $\frac{7}{x} = \frac{21}{9}$
By inspection
$x = 3$

5. $\frac{2x}{5} = \frac{16}{9}$
multiply by 45
$45 \cdot \frac{2x}{5} = 45 \cdot \frac{16}{9}$
$9 \cdot (2x) = 5 \cdot 16$
$18x = 80$
divide by 18
$x = \frac{80}{18} = \frac{40}{9} = 4.4\overline{4}$

7. $\frac{x+2}{5} = \frac{16}{10}$
multiply by 10
$10 \cdot \frac{x+2}{5} = 10 \cdot \frac{16}{10}$
$2(x + 2) = 16$
$2x + 4 = 16$
$\quad -4 \quad -4$
$2x = 12$
divide by 2
$x = 6$

9. $\frac{5}{12} = \frac{x+7.5}{42}$
multiply by 84
$84 \cdot \frac{5}{12} = 84 \cdot \frac{x+7.5}{42}$
$7 \cdot 5 = 2(x + 7.5)$
$2x + 15 = 35$
$\quad -15 \quad -15$
$2x = 20$
divide by 2
$x = 10$

11. $\frac{17}{36} = \frac{3}{2x+5}$
multiply by $36(2x + 5)$
$36(2x+5)\frac{17}{36} = 36(2x+5)\frac{3}{2x+5}$
$17(2x + 5) = 36 \cdot 3$
$34x + 85 = 108$
$\quad -85 \quad -85$
$34x = 23$
divide by 34
$x = \frac{23}{34} \approx 0.68$

13. $\frac{2.5 \text{ cups}}{48 \text{ cookies}} = \frac{x \text{ cups}}{84 \text{ cookies}}$
multiply by 336
$336 \cdot \frac{2.5}{48} = 336 \cdot \frac{x}{84}$
$7 \cdot 2.5 = 4x$
$4x = 17.5$
divide by 4
$x = 4.375$ cups

14. $\frac{1 \text{ gallon}}{440 \text{ ft}^2} = \frac{x \text{ gallon}}{1750 \text{ ft}^2}$
multiply by 77000
$77000 \cdot \frac{1}{440} = 77000 \cdot \frac{x}{1750}$
$175 = 44x$
$99x = 300$
divide by 44
$x = \frac{175}{44} \approx 3.98$ gallons
$3\frac{1}{2}$ gallons is not enough paint

Lesson 8 How Far Is It?

Homework

1. $\frac{1\text{ in}}{250\text{ mi}} = \frac{1.5\text{ in}}{x\text{ mi}}$
   multiply both sides by $250x$
   $250 \cdot x \frac{1}{250} = 250 \cdot x \frac{1.5}{x}$
   $x = 250 \cdot 1.5 = 375$ miles

2. $\frac{1\text{ in}}{250\text{ mi}} = \frac{x\text{ in}}{2805\text{ mi}}$
   multiply both sides by $250 \cdot 2805$
   $250 \cdot 2805 \frac{1}{250} = 250 \cdot 2805 \frac{x}{2805}$
   $250x = 2805$
   divide by 250
   $x = \frac{2805}{250} \approx 11.2$ inches

7. $C = 2\pi r \approx 2 \cdot 3.14 \cdot (8\text{ inches}) = 50.24$ inches

Lesson 9 How Big Is It?

Homework

2. Area $= 2 \cdot 6 = 12$ square units

4. Area $= \pi \cdot (4.5)^2 = 20.25\pi \approx 63.62$ square units

7. Area = Area of semicircle + area of rectangle $= \frac{1}{2}\pi \cdot (2.5)^2 + 5 \cdot 7 = \frac{25}{8}\pi + 35 \approx 74.3$ square units

Lesson 10 Fitting a Square Die onto a Round Wafer

1. Going from a 4-inch wafer to a 6-inch wafer increases the usable area and therefore increases the yield of good die for a given level of contamination expressed in particles per square inch.

3. Given that each good die produces a revenue of \$300, the total revenue for each wafer with no contamination is:
   a. \$9,600
   b. \$26,400
   c. \$49,200
   d. \$117,600

Lesson 11 How Much Does It Hold?

Homework

1. Given the radius of the earth to be approximately 20,926,435 feet, the volume of the earth is
   $V = \frac{4}{3}\pi r^3 \approx \frac{4}{3}\pi(20{,}926{,}435)^3 \approx 3.8386 \times 10^{22}$ cubic feet.

Linear Behavior

Lesson 1 Costs of Higher Education

Homework

1. Private College Tuition data

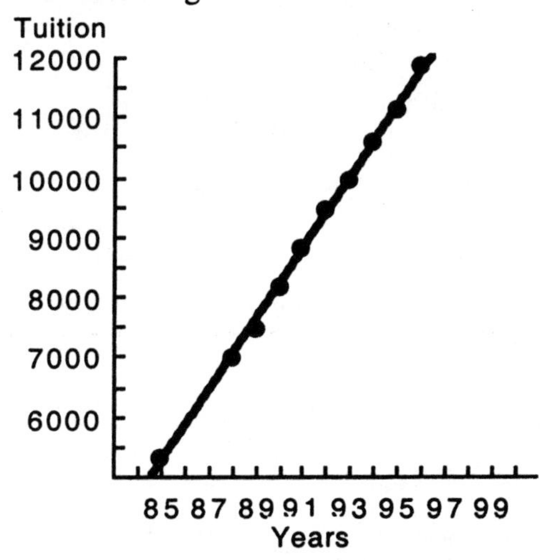

Tuition and Fees for Private Colleges

5. Choose b. All data the points are close, rather than being exactly on a few points and far from the others.

6. a.

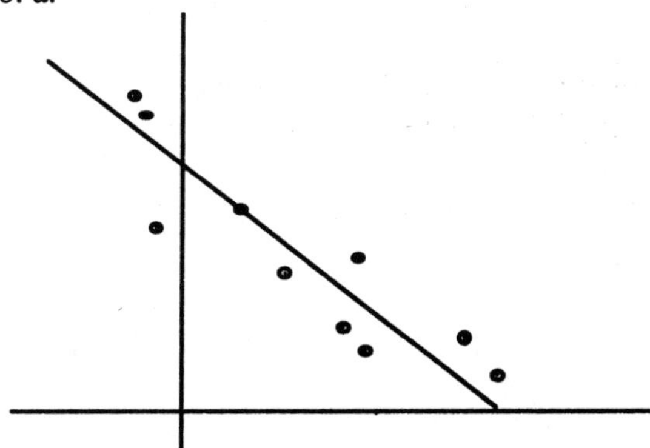

b.

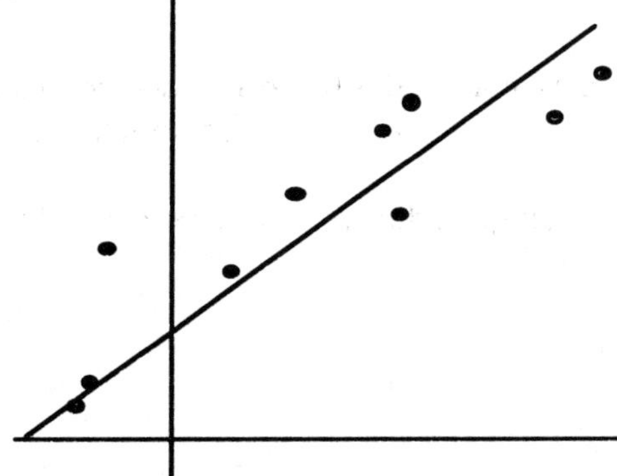

Lesson 2 The Price is Going Up

Homework

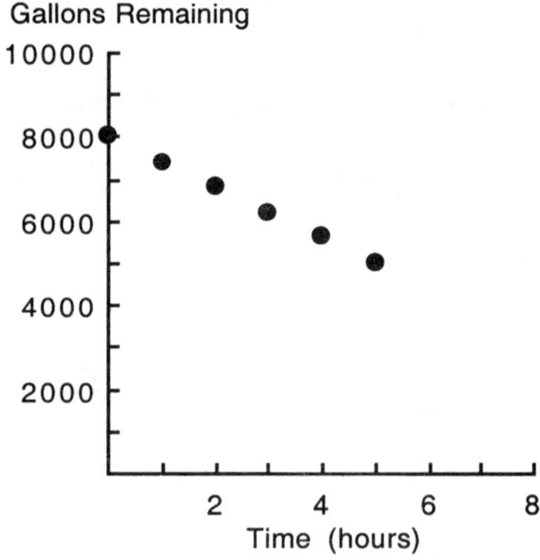

1. Plot the data:

   a. $m = \frac{8000-5000}{0-5} = \frac{3000}{-5} = -600$
      The number of gallons is decreasing at the rate of 600 gallons per hour.

   b. The y-intercept, (0, 8000) represents the starting point of the flight, a full tank.

   c. After 10 hours., there will be $8000 - 10 \cdot 600 = 2000$ gallons of fuel remaining.

   d. The plane will run out of gas after 13 hours and 20 minutes. After 13 hours there is only 200 gallons left. That leaves only 1/3 of an hour of fuel remaining.

   e. The amount of fuel remaining after each hour represents arithmetic growth (actually arithmetic decay) at a constant rate of –600 gallons per hour.

   f. Answers will vary, but should include wind speed, wind direction, altitude, take-offs and landings.

4. a. m = \$70 per year
   b. m = 74 bricks per layer
   c. m = –\$166.67 per year or price decreases \$166.67 per year
   d. m = –250 people per year or population is decreasing at the rate of 250 people per year

Lesson 3 Body Graphing

Homework

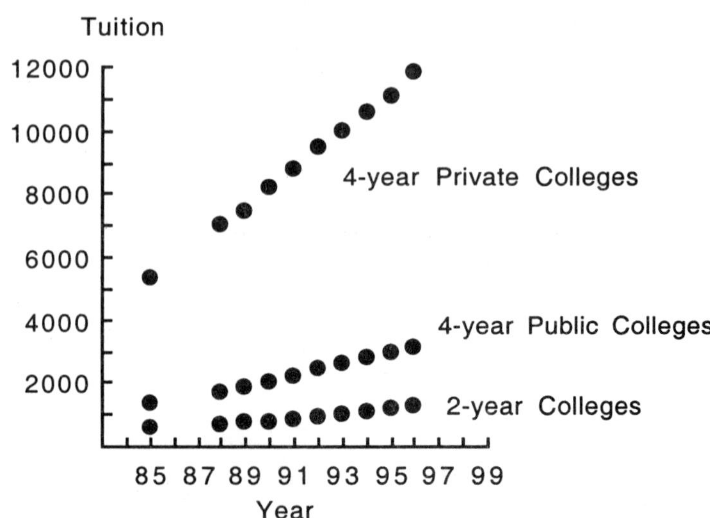

1. Answers will vary but should be close to:
   2-year College slope = \$60/yr
   4-year College slope = \$160/yr
   Private College slope = \$595/yr

   Note that the larger the value of the slope the steeper the line.

6. a., b. see graph

   c. $m = \dfrac{9-3}{3-0} = \dfrac{6}{3} = 2$

   d. $m = \dfrac{4-0}{3-1} = \dfrac{4}{2} = 2$

   e. The lines are parallel and the slopes are equal.

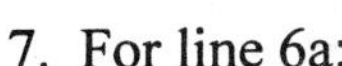

7. For line 6a:

   a. x-intercept = (–1.5, 0)

   b. y-intercept = (0, 3)

   For line 6b:

   c. x-intercept = (1, 0)

   d. y-intercept = (0, –2)

9. a. i.

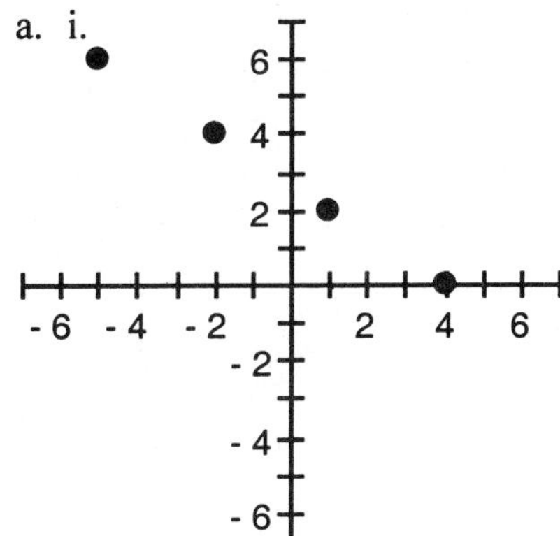

ii.

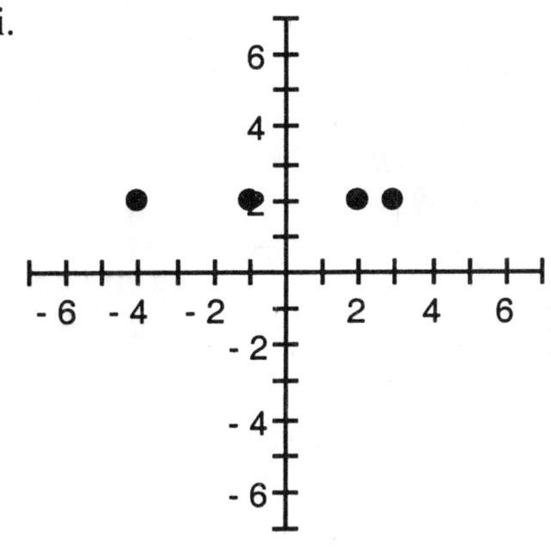

iii.

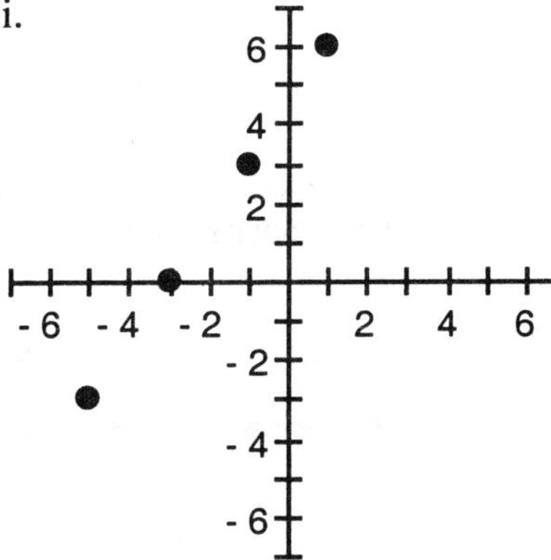

iv.

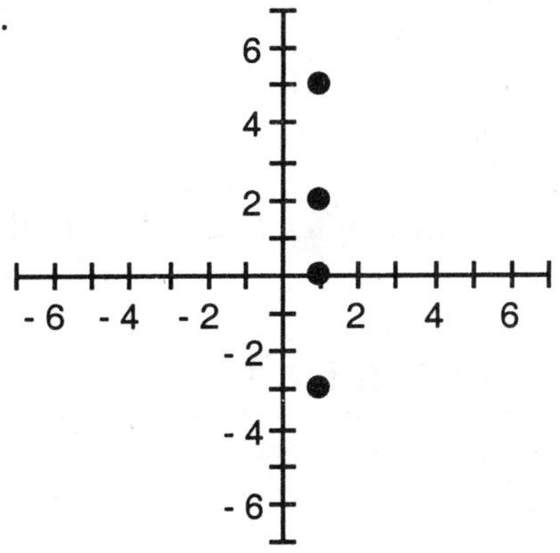

b. i. $m = \dfrac{2-0}{1-4} = -\dfrac{2}{3}$ ii. $m = \dfrac{2-2}{3-2} = 0$ iii. $m = \dfrac{6-3}{1-(-1)} = \dfrac{3}{2}$

iv. $m = \dfrac{5-2}{1-1} = \dfrac{3}{0}$, undefined

c. i. The line drops 2 units for every horizontal change of 3 to the right.

ii. The line is a horizontal line.

iii. The line rises 3 units for every change of 2 units to the right.

iv. The line is a vertical line.

Lesson 4 A Business Venture

Homework

5. a. -1 b. $\frac{1}{2}$ c. $\frac{2}{3}$ d. 0

6. a. $-\frac{1}{10}$ b. $\frac{1}{20}$ c. $\frac{2000}{3}$ d. undefined e. $\frac{23}{2200}$ f. $-\frac{9}{125}$

7. A grade of 6% is the same as a slope of $\frac{6}{100}$. This means that for every 100 feet of run there is a 6-foot rise or drop.

8. a., b.

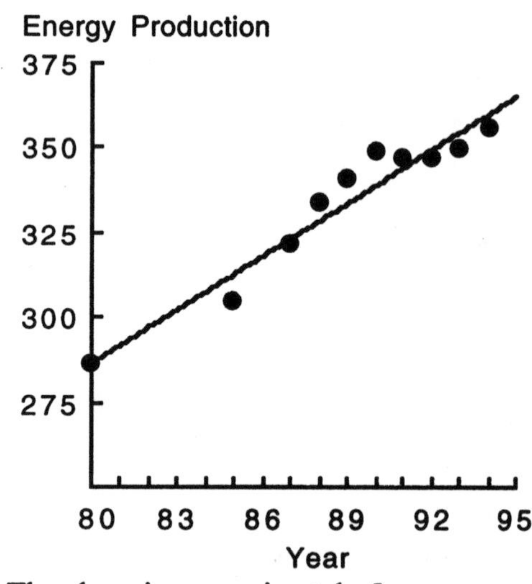

c. The slope is approximately 5.

d. Energy production increased at the rate of 5 quadrillion Btu per year. The linear regression slope is 5.23.

Lesson 5 The Bouncing Ball

Homework

1. The ordered pair (45, 89) represents that student 5 studied for 45 minutes and earned a score of 89 on the test.

4. The y-intercept is near (0, 37).

6. A student who studies 55 minutes should score near 100.

Lesson 6 Shrimp Dinners and Input/Output Machines

Homework

1. a.

Input	Output
3	18.8
0	–2.2
0.25	–0.45
2	11.8
0.75	3.05

b. 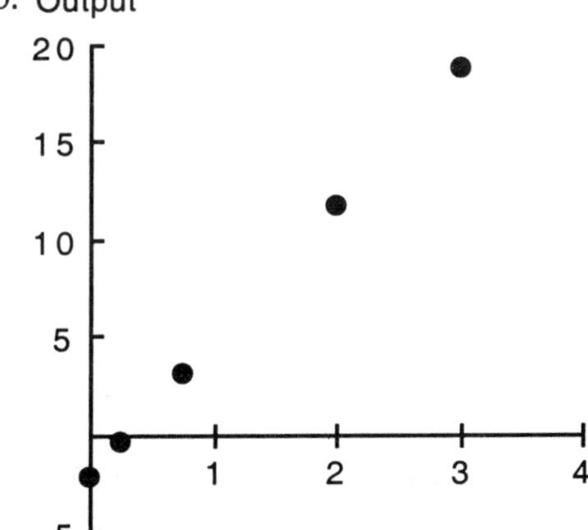

c. The y-intercept is (0, –2.2) and the slope is $\dfrac{18.8-11.8}{3-2}=\dfrac{7}{1}=7$.

d. $y=7x-2.2$

3. a. 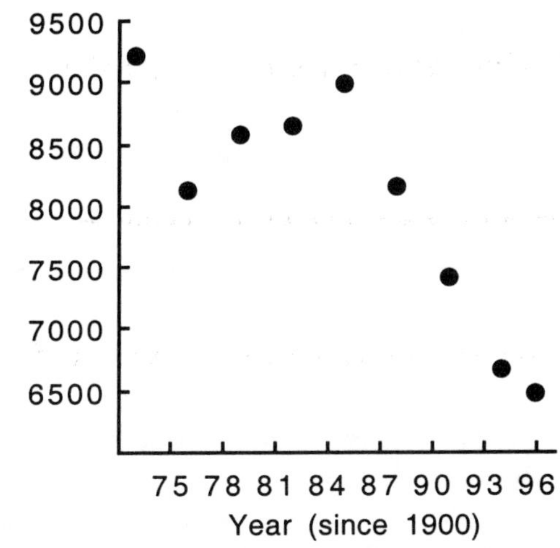

b. The slope depends on the points that are selected. For example, using (73, 9208) and (91, 7417), the slope is $\dfrac{7417-9208}{91-73}=\dfrac{-1791}{18}=-99.5$. The negative sign indicates that production is decreasing. Production is declining at the rate of 99.5 thousand barrels per day.

c. The input is the year and the output is daily production of crude oil in thousands of barrels.

Lesson 7 Driving Us Crazy

Homework

2. $m = 0.5$
   $b = 2$
   $y = 0.5x + 2$

4. $m = -1$
   $b = -1$
   $y = -x - 1$

6. $m = 0.5$
   $b = -2$
   $y = 0.5x - 2$

9. Let d = distance traveled and t = time in hours, then $d = 45t$.

Lesson 8 Get That Degree!

Homework

3. In 1998, a high school graduate would earn \$14,000, an associate degree holder would earn \$36,000 and someone with a bachelor's would earn \$40,000.

5. For the high school data, use (0, 10885) and (5, 12539) to get the slope $\frac{12539-10885}{5-0} \approx 331$. For the associate degree data, use (0, 19281) and (5, 27771) to get the slope $\frac{27771-19281}{5-0} = 1698$. For the bachelor degree data, use (0, 23722) and (5, 32152) to get the slope $\frac{32152-23722}{5-0} = 1686$.

7. In 1998, using the equations, a high school graduate would earn \$14,176.91, an associate degree holder would earn \$36,041.67 and someone with a bachelor's would earn \$40,40.33.

12. a. Let TS = test score and MS = minutes studied, then TS = 0.87·MS + 47.6.
    b. Answers will vary.
    c. You increase your score 0.87 point for each minute that you study.
    d. The y-intercept could be interpreted to mean if you do not study, your score should be a 48.

Lesson 9 Who Gives a Chirp?

Homework

3. Using t = years since 1983, the equation is $y = 48095 \cdot t + 390416$, where y = cost of a commercial.
   Using t = years since 1980, the equation is $y = 48095 \cdot t + 246131$, where y = cost of a commercial.
   Using t = years since 1900, the equation is $y = 48095 \cdot t - 3601488$, where y = cost of a commercial.
   Using t = year, the equation is $y = 48095 \cdot t - 94982440$, where y = cost of a commercial.

5. In 1996, the cost of a commercial would be \$1,014,000 to \$1,016,000 depending on rounding. In 2000, a commercial would cost \$1,206,000 to \$1,208,000.

7. With a budget of $10 million in 2000, you could buy 8 ads with money left over.

Lesson 10 "It Fits!" Story Time

Homework

1. Choose *a*. Maximum height gradually increases the longer the child swings. Choice *d* can be justified if the child is held and then pushed.

3. Choose *b*. The amount of money left gradually declines to 0.

5. Answers will vary, one possibility is shown:

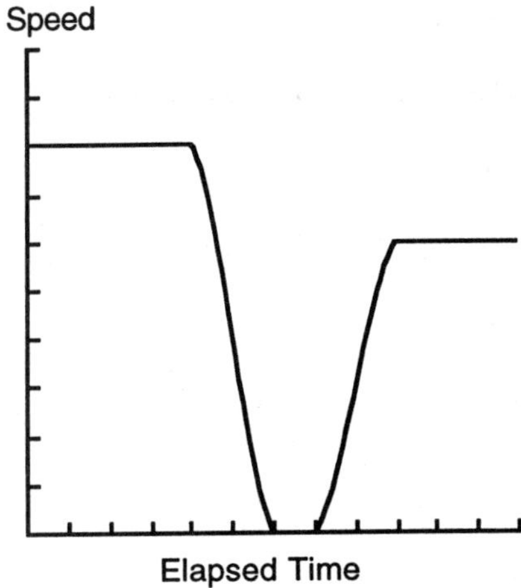

Nonlinear Behavior

Lesson 1 Quadratic Functions and Parabolas

Nitty Gritty: Solving for Any Variable in an Equation

1. $5y = 12$

$y = \frac{12}{5}$

3. $3x + 1 = 19$

$3x + 1 - 1 = 19 - 1$

$\frac{3x}{3} = \frac{18}{3}$

$x = 6$

5. $3a = 2b - 5$

$\frac{3a}{3} = \frac{2b - 5}{3}$

$a = \frac{2}{3}b - \frac{5}{3}$

7. $9x + 4y = 10$

$9x - 9x + 4y = 10 - 9x$

$4y = 10 - 9x$

$y = \frac{10}{4} - \frac{9}{4}x$

$y = \frac{5}{2} - \frac{9}{4}x$

9. $2x + 3y = 5x - 4y$

$2x - 2x + 3y + 4y = 5x - 2x - 4y + 4y$

$7y = 3x$

$y = \frac{3}{7}x$

11. $5 = \frac{3a}{4}$

$4 \cdot 5 = \frac{3a}{4} \cdot 4$

$20 = 3a$

$\frac{20}{3} = a$

13. $c = \frac{2 + 3y}{4}$

$4c = 2 + 3y$

$4c - 2 = 2 - 2 + 3y$

$4c - 2 = 3y$

$\frac{4}{3}c - \frac{2}{3} = y$

Homework

4. a. $y = -x^2 + 3x - 5$
   $a = -1$ so the parabola is concave down. The parabola increases then decreases.
   $c = -5$ means that the y-intercept is at (0,–5).
   Because b is positive and a is negative, the graph is shifted up and to the right of the origin.

   b. $y = 3x^2 - 3x - 5$
   $a = 3$ means that the parabola is concave up. It decreases, then increases. It is narrow.
   $c = -5$ means that the y-intercept is at (0,–5).
   Because b is negative and a is positive, the parabola is shifted down and to the right of the origin.

7.
   a.

X	Y	1st Difference	2nd Difference
1	1		
		–1	
2	0		2
		1	
3	1		2
		3	
4	4		2
		5	
5	9		

The data are quadratic because the second differences are the same.

Lesson 2 Products and Factoring

Homework

1. a. It is concave up. It has two x-intercepts, one positive and the other negative. Its y-intercept is negative. It looks rather narrow.

2. a. y_1 is positive when $x > 2$
   e. The parabola passes through the same x-intercepts as the two linear graphs.

3. c. The functions differ in sign when $-1 < x < 3$

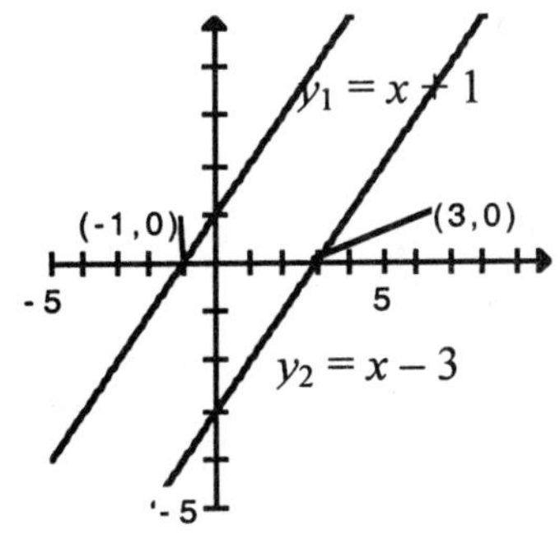

5. a. x-intercepts: (–1,0), (–2,0)
   $y = (x + 1)(x + 2)$

6. a. $(x-6)(x+1)$

7. a. $x^2 + 3x + 2 = 0$ can be solved in two ways.

Factor $x^2 + 3x + 2 = 0$
$(x+2)(x+1) = 0$
$x + 2 = 0$ or $x + 1 = 0$
$x = -1, x = -2$

Graph: $y = x^2 + 3x + 2$
Notice that the x-intercepts are at $x = -2$ and $x = -1$.
That is the solution to the equation.

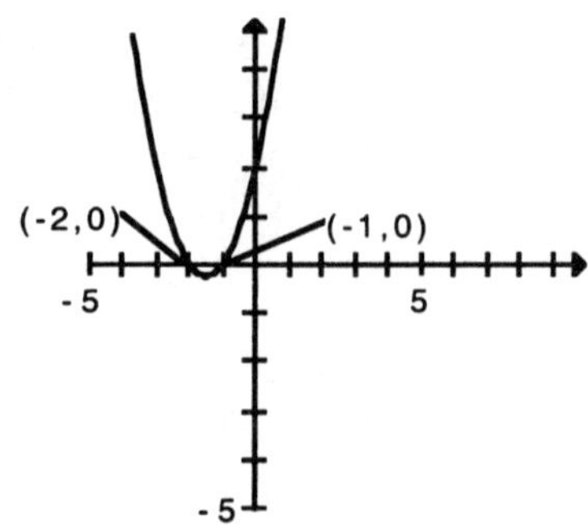

8. There are many ways to solve these equations. Here are two.

a. Algebraically
Get a zero on one side.
Factor to solve.
$x^2 - 5x = 6$
$x^2 - 5x - 6 = 0$
$(x-6)(x+1) = 0$
$x - 6 = 0$ or $x + 1 = 0$
$x = 6, x = -1$

Graphically
Graph the expression on each side of the equal sign.
$y_1 = x^2 - 5x$
$y_2 = 6$
The point(s) where the two graphs intersect are $(-6,6)$ and $(6,6)$
Since we want the x-values, the solution to the equation is $x = -1$ or $x = 6$.

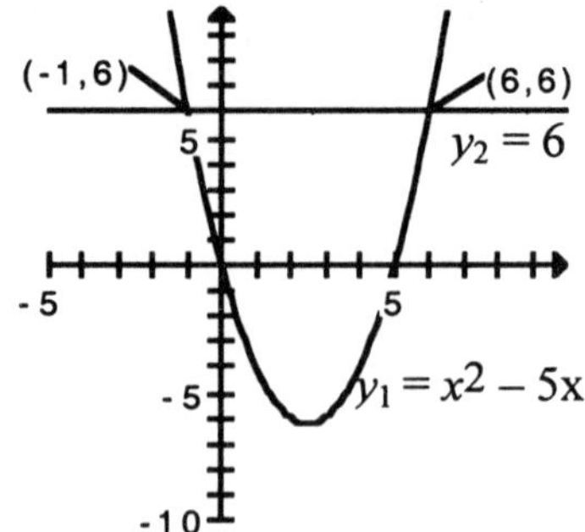

Lesson 3 Solving Equations

Homework

1. $-3x^2 = 2x - 5$
$-3x^2 - 2x + 5 = 2x - 5 - 2x + 5$
$-3x^2 - 2x + 5 = 0$
$a = -3, b = -2, c = 5$

$$x = \frac{-(-2) \pm \sqrt{(-2)^2 - 4(-3)(5)}}{2 \cdot (-3)}$$

$$x = \frac{2 \pm \sqrt{4 + 60}}{-6} = \frac{2 \pm \sqrt{64}}{-6}$$

$$x = \frac{2 \pm 8}{-6}$$

$$x = \frac{2+8}{-6} \text{ or } x = \frac{2-8}{-6}$$

$$x = -\frac{5}{3} \text{ or } x = 1$$

or

$-3x^2 = 2x - 5$
$-3x^2 + 3x^2 = 3x^2 + 2x - 5$
$0 = 3x^2 + 2x - 5$
$a = 3, b = 2, c = -5$

gives the same result

$$x = -\frac{5}{3} \text{ or } x = 1$$

Graphically, let $y_1 = -3x^2$ and $y_2 = 2x - 5$; then find the x-values of the points of intersection.

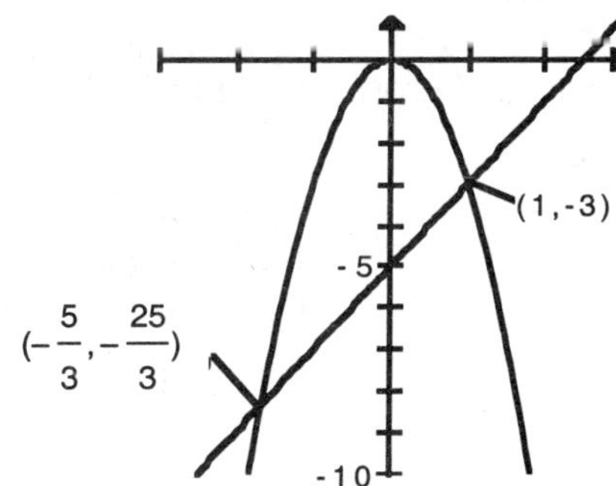

With the Quadratic Formula programmed into your calculator, enter $a = 3$, $b = 2$, $c = -5$ to get the result $x = 1$ or $x = -\frac{5}{3}$.

4. $x^2 + 3x - 5 = 0$

Using the quadratic formula, $a = 1$, $b = 3$, $c = -5$

$$x = \frac{-3 \pm \sqrt{3^2 \pm 4(1)(\pm 5)}}{2 \cdot 1}$$

$$x = \frac{-3 \pm \sqrt{9 + 20}}{2}$$

$$x = \frac{-3 \pm \sqrt{29}}{2}$$

$$x = \frac{-3 + \sqrt{29}}{2} \text{ or } x = \frac{-3 \pm \sqrt{29}}{2}$$

$x \approx 1.19$ or $x \approx -4.19$

Using the programmable calculator method, $a = 1$, $b = 3$, $c = -5$

$x \approx 1.19258$ or $x \approx -4.19258$

9. $-4.9x^2 + 5.6x + 120 = 0$ $\quad x \approx -4.41$ or 5.55

To use the quadratic formula, multiply both sides of the equation by 10, so each coefficient is an integer.

$-49x^2 + 56x + 1200 = 0$

$a = -49$, $b = 56$, $c = 1200$

$$x = \frac{-56 \pm \sqrt{56^2 - 4(-49)(1200)}}{2 \cdot (-49)}$$

$$x = \frac{-56 \pm \sqrt{238{,}336}}{-98}$$

$x \approx -4.41$ or $x \approx 5.55$

By graphing,

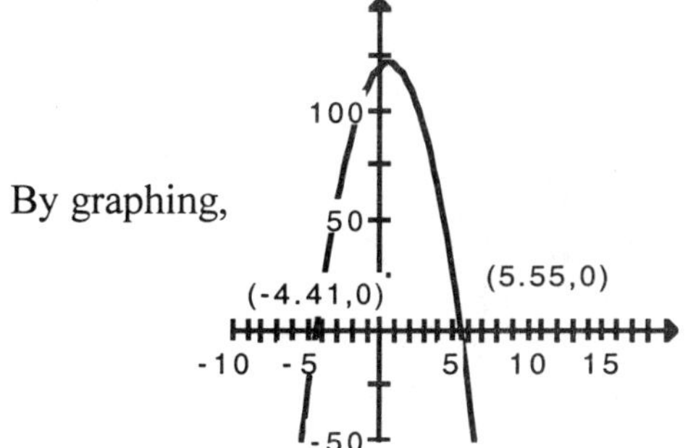

10. Since we start at a height of 415 feet, we know that $h_0 = 415$. The equation is

$h(t) = -16t^2 + 415$

We want to know how long it takes for $h(t)$ to be zero, at ground level. So, we solve

$0 = -16t^2 + 415$

By the square root method (since $b = 0$)

$16t^2 = 415$

$$t^2 = \frac{415}{16}$$

$$\sqrt{t^2} = \pm\sqrt{\frac{415}{16}}$$

$$t \approx \pm\sqrt{25.9375}$$

$t \approx \pm 5.09$

Using the quadratic formula, $a = -16$, $b = 0$, $c = 415$

$$t = \frac{-0 \pm \sqrt{0^2 - 4(-16)(415)}}{2\cdot(-16)}$$

$$t = \frac{\pm\sqrt{26{,}560}}{-32}$$

$t \approx \pm 5.09$

It takes about 5 seconds to drop the 415 feet on the Superman ride at Magic Mountain.

Lesson 4 Complex Numbers

Nitty Gritty: Operations on Complex Numbers

1. $(4 + 2i) + (3 + i)$
   $4 + 3 + 2i + i$
   $7 + 3i$

3. $(0 + 3i) + (5 + 0i)$
   $0 + 5 + 3i + 0i$
   $5 + 3i$

5. $(3 + 3i) - (2 + 2i)$
   $3 - 2 + 3i - 2i$
   $1 + i$

7. $(-1 + 4i) - (4 + i)$
   $-1 - 4 + 4i - i$
   $-5 + 3i$

9. $(4 + 2i)(3 + i)$
   $4\cdot3 + 4i + 3\cdot2i + 2i\cdot i$
   $12 + 4i + 6i + 2i^2$
   $12 + 10i + 2(-1)$
   $12 - 2 + 10i$
   $10 + 10i$

11. $(0 + 3i)(5 + 0i)$
   $0 + 0i + 15i + 0$
   $15i$

13. $\dfrac{4+2i}{3+i}$

$\dfrac{4+2i}{3+i}\cdot\dfrac{3-i}{3-i}$

$\dfrac{12-4i+6i-12i^2}{9-3i+3i-i^2}$

$\dfrac{12+2-4i+6i}{9+1-3i+3i}$

$\dfrac{14+2i}{10}$

$\dfrac{7}{5}+\dfrac{1}{5}i$ or $1.4+0.2i$

17. $(-2+5i)[(3+2i)-(5-3i)]$

$(-2+5i)[3-5+2i-(-3i)]$

$(-2+5i)(-2+5i)$

$4-10i-10i+25i^2$

$4-25-20i$

$-21-20i$

15. $\dfrac{0-3i}{5+0i}$

$-\dfrac{3i}{5}$

or $-0.6i$

Homework

1. $f(x)=x^2+3x+5$

Since this graph does not cross the x-axis, $f(x)$ has not real roots. It has two complex roots.

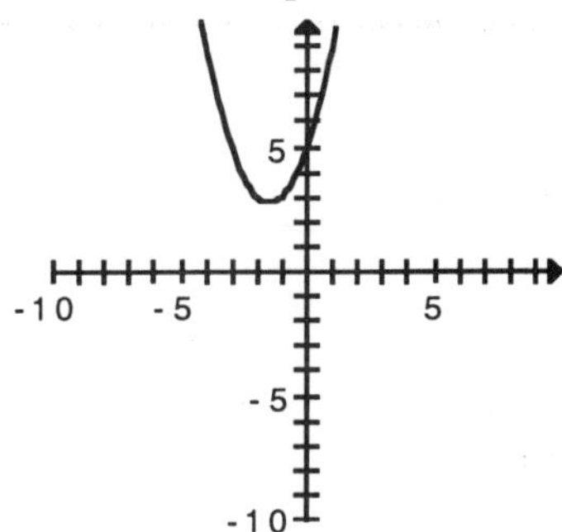

4. $f(x)=x^2-4x+4$

This graph has one real root because it touches the x-axis in only one point.

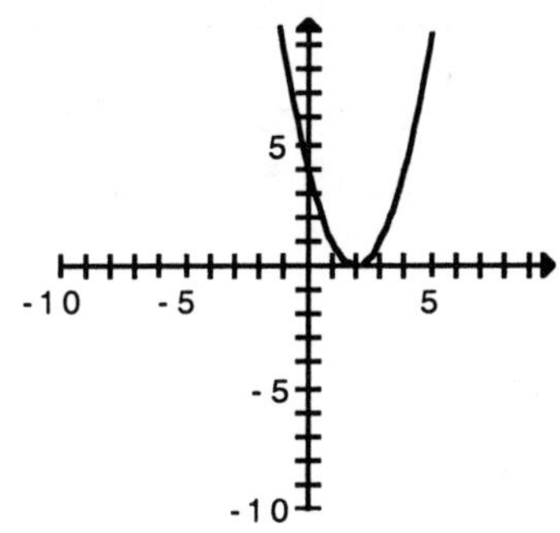

5. $x^2+3x+1=0$

Using the quadratic formula,

$a=1,\ b=3,\ c=1$

$$x=\frac{-3\pm\sqrt{3^2-4(1)(1)}}{2\cdot 1}$$

$$x=\frac{-3\pm\sqrt{9-4}}{2}$$

$$x=\frac{-3+\sqrt{5}}{2}\ \text{or}\ x=\frac{-3-\sqrt{5}}{2}$$

$x\approx 0.382$ or $x\approx -2.618$

6. $-4x^2-2x-4=0$

$-2(2x^2+x+2)=0$

$2x^2+x+2=0$

$a=2,\ b=1,\ c=2$

$$x=\frac{-1\pm\sqrt{1^2-4(2)(2)}}{2\cdot 2}$$

$$x=\frac{-1\pm\sqrt{-15}}{4}$$

$$x=-\frac{1}{4}+\frac{\sqrt{15}}{4}i\ \text{or}\ x=-\frac{1}{4}-\frac{\sqrt{15}}{4}i$$

9. $0.15x^2 + 1.2x - 4.7 = 0$

$a = 0.15,\ b = 1.2,\ c = -4.7$

$$x = \frac{-1.2 \pm \sqrt{(1.2)^2 - 4(0.15)(-4.7)}}{2 \cdot (0.15)}$$

$$x = \frac{-1.2 \pm \sqrt{4.26}}{0.30}$$

$x \approx 2.880$ or ≈ -10.8805

12. a. Using $I = \dfrac{V}{Z}$,

$$6 = \frac{15 + 18i}{Z}$$

$$6Z = 15 + 18i$$

$Z = \dfrac{15 + 18i}{6}$. The impedance is $\dfrac{5}{2} + 3i$ ohms

Lesson 5 Cubic Graphs and Equations

Homework

1. b. The graph is generally decreasing, so the x^3 term must be negative. In addition, the y-intercept is (0, –6). Therefore, of the given choices, the equation must be $y = -x^3 + 2x^2 + 5x - 6$.
   c. The equation has four terms. The first term has a coefficient of –1. The fourth term is also negative.
   d. The graph is concave up and then concave down. It has a point of inflection when x is about 0.5. The graph crosses the x axis three times.
5. b. The regression equation is $y = 0.125x^3 - 2.625x^2 + 17.875x - 34.375$

Lesson 6 Modeling Cubic Data

Homework

1. The cubic regression equation for the MetLife data is $y = -0.002985x^3 + 0.6x^2 - 37.2x + 835$.

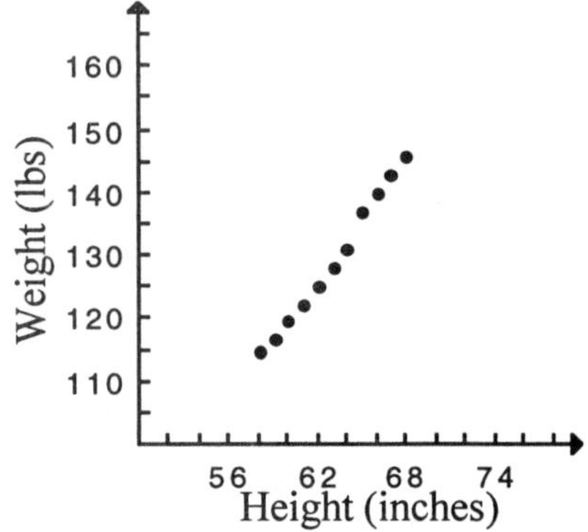

4. The only x-intercept is approximately $(-1.27, 0)$.

x	y
−2	−5.9
−1	1.45
0	4
1	3.55
2	1.9
3	0.85
4	2.2
5	7.75

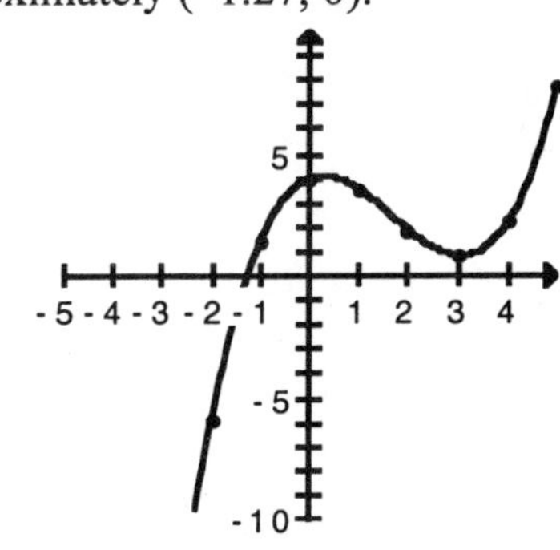

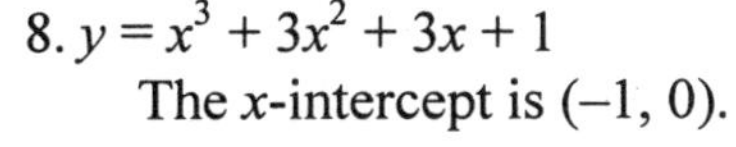

8. $y = x^3 + 3x^2 + 3x + 1$
   The x-intercept is $(-1, 0)$.

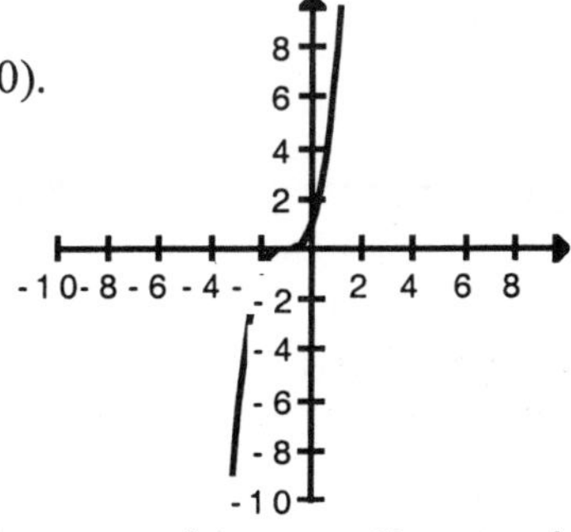

12. A cubic function always has one y-intercept. Because it is a function it can't have more than one y-intercept, else it would fail the vertical line test.

14. The volume (V) of a sphere: $V(r) = \frac{4}{3}\pi r^3$. To find the volume of a sphere with radius 3 feet,

    calculate $V(3) = \frac{4}{3}\pi(3)^3 = 36\pi$ cubic centimeters

    The volume of the sphere is about 113 cubic centimeters.

Lesson 7 Patterns in Cubic Data

Homework

2. These data are cubic because the third differences are the same.

X	Y	first differences	second differences	third differences
−6	23.8			
		−16.8		
−5	7		11.6	
		−5.2		−6
−4	1.8		5.6	
		0.4		−6
−3	2.2		−0.4	
		0		−6
−2	2.2		−6.4	
		−6.4		−6
−1	−4.2		−12.4	
		−18.8		
0	−23			

3. These data are quadratic because the second differences are the same.

X	Y	first differences	second differences
12	13		
		−12	
14	1		8
		−4	
16	−3		8
		4	
18	1		8
		12	
20	13		

Lesson 8 Radical Functions

Homework

1. $A = \frac{1}{2} \cdot a \cdot \sqrt{c^2 - a^2}$.

3. Person 1 has a mass of 50 kg has $L_1 = \sqrt[3]{50}$.

   Person 2 has a mass of 100 kg has $L_2 = \sqrt[3]{100}$.

$$\frac{\text{oxygen consumption for person 2}}{\text{oxygen consumption for person 1}} = \frac{L_2^{\,2}}{L_1^{\,2}}$$

$$\frac{\text{oxygen consumption for person 2}}{250} = \frac{\left(\sqrt[3]{100}\right)^2}{\left(\sqrt[3]{50}\right)^2}$$

$$\text{oxygen consumption for person 2} = 250 \cdot \left(\sqrt[3]{\frac{100}{50}}\right)^2$$

$$\text{oxygen consumption for person 2} = 250 \cdot \left(\sqrt[3]{2}\right)^2$$

oxygen consumption for person 2 $= 250 \cdot \sqrt[3]{4}$ or about 396.85 cc/min

5. $$\frac{\text{rate of blood pumped for person 2}}{\text{rate of blood pumped for person 1}} = \frac{L_2^{\,2}}{L_1^{\,2}}$$

$$\frac{\text{rate of blood pumped for person 2}}{6} = \frac{\sqrt[3]{100}^{\,2}}{\sqrt[3]{50}^{\,2}}$$

rate of blood pumped for person 2 $= 6 \cdot \sqrt[3]{2}$ or about 9.52 liters/min.

7. To find an expression that gives the measure of the radius of a sphere as a function of volume, isolate r, the radius.

$V = \frac{4}{3} \cdot \pi \cdot r^3$

$\frac{3}{4}V = \frac{3}{4} \cdot \frac{4}{3} \cdot \pi \cdot r^3$ multiply both sides by $\frac{3}{4}$

$\frac{3V}{4\pi} = \frac{\pi \cdot r^3}{\pi}$ divide both sides by π

$\sqrt[3]{\frac{3V}{4\pi}} = \sqrt[3]{r^3}$ cube root both sides

$\sqrt[3]{\frac{3V}{4\pi}} = r$ We're done. We have solved for r in terms of V.

12. This problem can be done as shown in problem 11. Here's another way to go about it.

$A = \frac{1}{2}bh$ for any triangle. Since $A = 77$ and $b = 7$, we have

$77 = \frac{1}{2}3 \cdot h$

$154 = 3 \cdot h$

$51.3333 \approx h$

c h 3

We now know $h \approx 51.3333$ and $b = 3$. We have enough to use the Pythagorean Theorem to find the hypotenuse, c.

$a^2 + b^2 = c^2$

$51.3333^2 + 3^2 \approx c^2$ we use ≈ because we've already rounded the height.

$2635.11 + 9 \approx c^2$

$2644.11 \approx c^2$

$\pm\sqrt{2644.11} \approx \sqrt{c^2}$

$51.42 \approx c$ choose the positive square root

13. We can use the formula from Activity 2.

$\text{Perimeter} = x + b + \sqrt{x^2 - b^2}$

$76 = x + 6 + \sqrt{x^2 - 6^2}$

$76 - 6 - x = \sqrt{x^2 - 36}$

$70 - x = \sqrt{x^2 - 36}$

$(70 - x)^2 = \left(\sqrt{x^2 - 36}\right)^2$

$4900 - 140x + x^2 = x^2 - 36$

$-140x = -4936$

$x \approx 35.26$

Lesson 9 Absolute Value

Homework

1. Since 10% of $1\frac{5}{8}$ is 0.1625, we can write $\left|x-1\frac{5}{8}\right| < 0.1625$.

2. The smallest the beam could be is 10 times $(1\frac{5}{8} - 0.1625)$ which is 14.625 inches.

   The largest the beam could be is 10 times $(1\frac{5}{8} + 0.1625)$ which is 17.875 inches.

   The beam could range between 14.625 inches and 17.875 inches.

5. Since the horizontal line is $y = 2$, we need an inequality with a 2 on one side. This eliminates a, c, and f. The slanted line appears to have a y-intercept of $(0, -3)$ and a slope of 1. Thus, a possible equation is $y = x - 3$. Therefore, it looks like b and d could be represented by the graph.

6. $x + 3 < 9$
   $x + 3 - 3 > 9 - 3$
   $x < 6$

8. $5 - x < 3$
   $5 - 5 - x < 3 - 5$
   $-x < -2$
   $(-1)(-x) > (-1)(-2)$
   $x > 2$

12. $|x-3| < 1$
    $-1 < x - 3 < 1$
    $2 < x < 4$

16. $|3x-10| < 4$
    $-4 < 3x - 10 < 4$
    $6 < 3x < 14$
    $2 < x < \frac{14}{3}$

18. $|x-10| = 1.5$
    By graphing, $y_1 = |x-10|$ and $y_2 = 1.5$

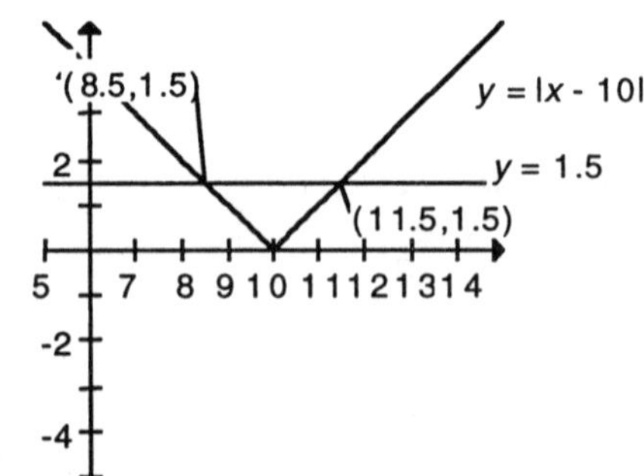

The two graphs are equal at $x = 8.5$ and at $x = 11.5$

21. $|3x + 2| = 7$

    We understand from this equation that the quantity $3x + 2$ must be equal to 7 or to -7. Subtract 2 from each of those values, revealing that $3x$ has to be equal to 5 or -9. Dividing by 3, shows that x is either $\frac{5}{3}$ or -3.

Lesson 10 Rational Functions

Nitty Gritty: Solving Rational Equations Algebraically

1. $\dfrac{7x}{100 - x} = 25$

$7x = 25(100 - x)$

$7x = 2500 - 25x$

$32x = 2500$

$x = 78.125$

Check:

$\dfrac{7(78.125)}{100 - 78.125} \stackrel{?}{=} 25$

$\dfrac{546.875}{21.875}$

25

It checks.

3. $\dfrac{x}{x + 3} = 18$

$x = 18x + 54$

$-17x = 54$

$x = -\dfrac{54}{17}$

$x \approx -3.176$

Check:

$\dfrac{-3.176}{-3.176 + 3} \stackrel{?}{=} 18$

$\dfrac{-3.176}{-0.176}$

18.05. We'll call it good.

5. $\dfrac{x}{x + 2} + \dfrac{1}{x - 5} = 150$ — Multiply each term by $(x+2)(x-5)$

$x(x-5) + (x+2) = 150(x+2)(x-5)$ — Simplify

$x^2 - 5x + x + 2 = 150(x^2 - 3x - 10)$

$x^2 - 4x + 2 = 150x^2 - 450x - 1500$ — Collect like terms

$0 = 149x^2 - 446x - 1502$ — Set equal to zero

Use the quadratic formula, with $a = 149$, $b = -446$, and $c = -1502$

$$x = \frac{446 \pm \sqrt{(-446)^2 - 4(149)(-1502)}}{2(149)}$$

$$x = \frac{446 \pm \sqrt{1{,}094{,}108}}{298}$$

$x \approx 5.01$ or -2.01

Check:

$\dfrac{5.01}{5.01 + 2} + \dfrac{1}{5.01 - 5} \stackrel{?}{=} 150$

$0.715 + 100$

100.7 This is not very close to 150. Maybe it's because we rounded to two decimal places.

From the quadratic formula, record the answer to 4 decimal places.

$x \approx 5.0067$ or $x \approx -2.0134$

Check:

$\dfrac{5.0067}{5.0067 + 2} + \dfrac{1}{5.0067 - 5} \stackrel{?}{=} 150$

$0.7146 + 149.2537$

149.9683. That's better because it is quite close to 150. It checks.

So, the answers are $x \approx 5.0067$ or $x \approx -2.0134$.

7. $\dfrac{x}{x-5} + \dfrac{x}{x+2} = 8$ Multiply by $(x-5)(x+2)$

$x(x+2) + x(x-5) = 8(x-5)(x+2)$
$x^2 + 2x + x^2 - 5x = 8(x^2 - 3x - 10)$
$2x^2 - 3x = 8x^2 - 24x - 80$
$0 = 6x^2 - 21x - 80$
To solve this quadratic equation, graph $y = 6x^2 - 21x - 80$ and find the x-intercepts.

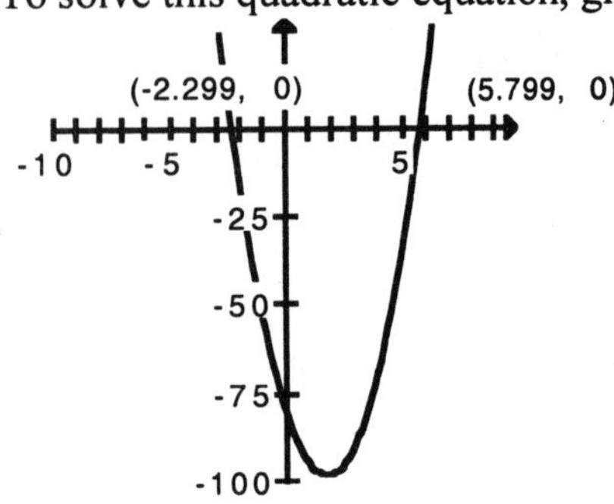

Rounding to two decimal places gives solutions of $x \approx 5.80$ or $x \approx -2.30$.
Check the proposed answer of –2.30:

$$\frac{-2.30}{-2.30-5} + \frac{-2.30}{-2.30+2} \stackrel{?}{=} 8$$

about 7.98. That's close to our target of 8. It checks.
Check the proposed answer of 5.80:

$$\frac{5.80}{5.80-5} + \frac{5.80}{5.80+2} \stackrel{?}{=} 8$$

about 7.99. That's close to our target of 8. It checks.
Note: rounding the values of the intercepts to 3 decimal places provides an even closer check.

9. $\dfrac{x}{x-3} = \dfrac{3}{x-3}$

Because the denominators are equal, the numerators must be equal.
$x = 3$

Check:

$$\frac{3}{3-3} \stackrel{?}{=} \frac{3}{3-3}$$

We get division by zero.
Therefore there is no solution.

11. $\dfrac{x}{x+2} = \dfrac{1}{2x-3}$

$x(2x-3) = x + 2,$
$x^2 - 2x - 1 = 0$
By the quadratic formula, we get
$x \approx 2.414$ or $x \approx -0.414$
Check:

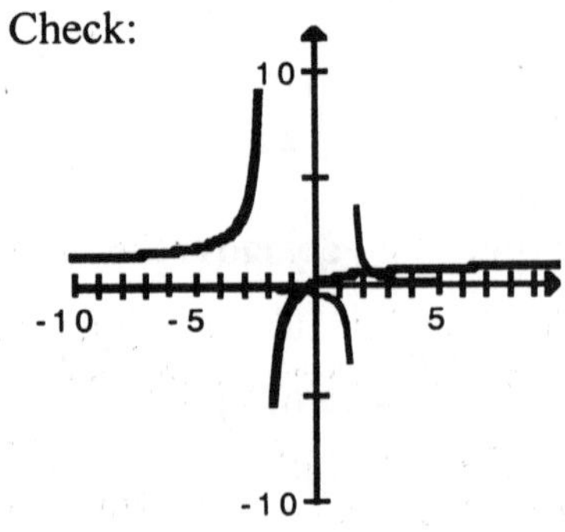

The two graphs intersect when
$x \approx 2.414$ or $x \approx -0.414$

Homework

1. For a. – d., find the point on the curve $y = \dfrac{28p}{100 - p}$ where y is 50 or 100 or 150 or 200. The corresponding p-values (along the horizontal axis) are:
   a. $p \approx 64.1$
   b. $p \approx 78.1$

3. $F(x) = \dfrac{40x}{10 + x}$

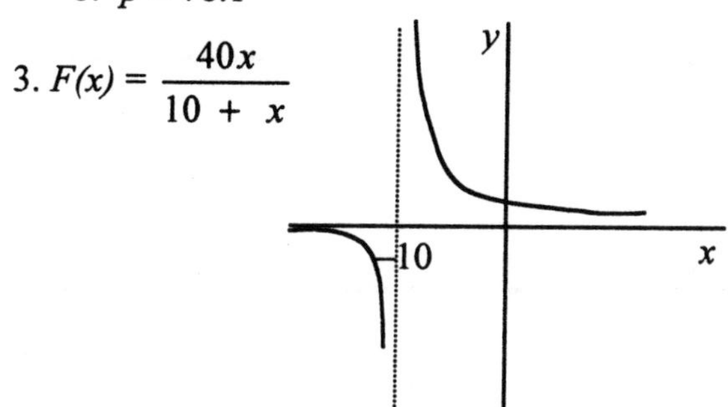

8. $f(p) = \dfrac{3p}{5 - p}$

   To identify the equation of a vertical asymptote, look for the value that creates a zero in the denominator. In this case 5 – 5 is zero so, $p = 5$ is the equation of the vertical asymptote.

Lesson 11 Sinusoidal Functions

Homework

2.

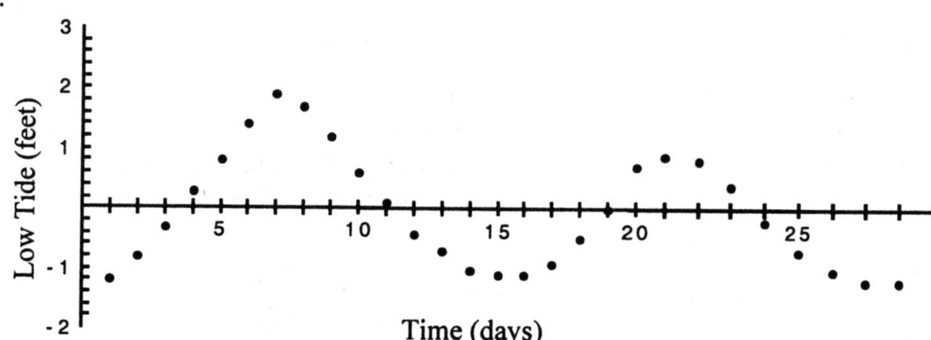

The regression equation for the lowest tides is given by $y = 1.21 \cdot \sin(0.456x - 1.81) - 0.035$

3. The height of the low tide on February 23rd. is estimated to be about 0.4 ft. if from the graph, 0.79 ft. if from the regression equation.

6. a. $3\sin x = 0.5$ at $x = 0.167$ or at $x = 2.97$
   c. $3\sin(2x - 1.57) = 0.5$ at $x = 0.869$, $x = 2.272$, $x = 4.01$, or $x = 5.41$

Patterns

Lesson 1 Introducing Patterns

none

Lesson 2 Introduction to Tessellations

Homework

1. a. Yes, these are regular hexagons.
   b. Yes, these are equilateral triangles.
   c. No, this shape is not regular since all sides and all angles are not equal.
   d. No, sides and angles are not equal.

Lesson 3 Angling for Patterns

Name	Number of Sides n	Angle-Measure m	Angle-Sum of the Polygon s
triangle	3	60°	180°
square	4	90°	360°
pentagon	5	108°	540°
hexagon	6	120°	720°
heptagon	7	≈128.57°	900°
octagon	8	135°	1080°

Activity 1–Finding a Formula for the Angle-Sum of a Polygon

8. $S = (n - 2) \cdot 180$

Homework

1.

Name	Number of Sides n	Angle-Measure m	Angle-Sum of the Polygon s
nonagon	9	140°	1260°
decagon	10	144°	1440°
dodecagon	12	150°	1800°
pentadecagon	15	156°	2340°
icosagon	20	162°	3240°

Lesson 4 Translating the Language of Tessellations

Homework

4. Translation and reflection

Lesson 5 Introduction to Polyhedra

✎To form a vertex, you need at least three faces. There is no maximum number.

Activity 1–How Many Regular Polyhedra Are There?

2.

Polygon	Number Meeting at Vertex	Angle-Sum (< 360°)
equilateral triangle	3	180°
equilateral triangle	4	240°
equilateral triangle	5	300°
square	3	270°
pentagon	3	324°

Homework

1.

Name	Number of Faces	Polygon Used	Angle-Sum of Vertex
tetrahedron	4	triangle	180°
hexahedron (cube)	6	square	270°
octahedron	8	triangle	240°
dodecahedron	12	pentagon	324°
icosahedron	20	triangle	300°

Lesson 6 Building Polyhedra

Homework

2. They are listed in increasing order of the number of faces.

3. Hexahedron, six seats

Lesson 7 Feeling a Bit Edgy

Name	Vertices	Edges	Faces
tetrahedron	4	6	4
cube	8	12	6
octahedron	6	12	8
dodecahedron	20	30	12
icosahedron	12	30	20

✎ $V - E + F = 2$

Lesson 8 Fractal Patterns

✎

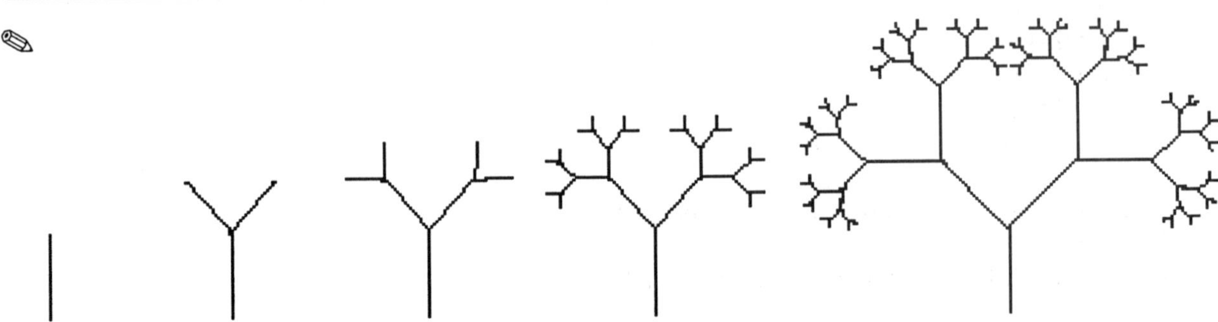

Lesson 9 Recursive Relations

Activity 1–Practicing with Recursive Relations

1. $B_1 = 2$,
   $B_2 = 0.5 \cdot 2 + 0.5 = 1.5$,
   $B_3 = 0.5 \cdot 1.5 + 0.5 = 1.25$,
   $B_4 = 0.5 \cdot 1.25 + 0.5 = 1.125$,
   $B_5 = 0.5 \cdot 1.125 + 0.5 = 1.0625$,
   $B_6 = 0.5 \cdot 1.0625 + 0.5 = 1.03125$,
   $B_7 = 0.5 \cdot 1.03125 + 0.5 = 1.015625$,
   $B_8 = 0.5 \cdot 1.015625 + 0.5 = 1.0078125$,
   $B_9 = 0.5 \cdot 1.0078125 + 0.5 = 1.00390625$,
   $B_{10} = 0.5 \cdot 1.00390625 + 0.5 = 1.001953125$

3. $B_1 = 4$,
   $B_2 = 0.5 \cdot 4 + 0.5 = 2.5$,
   $B_3 = 0.5 \cdot 2.5 + 0.5 = 1.75$,
   $B_4 = 0.5 \cdot 1.75 + 0.5 = 1.375$,
   $B_5 = 0.5 \cdot 1.375 + 0.5 = 1.1875$,
   $B_6 = 0.5 \cdot 1.1875 + 0.5 = 1.09375$,
   $B_7 = 0.5 \cdot 1.09375 + 0.5 = 1.046875$
   $B_8 = 0.5 \cdot 1.046875 + 0.5 = 1.0234375$,
   $B_9 = 0.5 \cdot 1.023435 + 0.5 = 1.01171875$,
   $B_{10} = 0.5 \cdot 1.01171875 + 0.5 = 1.005859375$

Homework

1. 2, 4, 6, 8, 10
   $a_1 = 2$
   $a_2 = 2 + 2 = 4$
   $a_3 = 4 + 2 = 6$
   $a_4 = 6 + 2 = 8$
   $a_5 = 8 + 2 = 10$

3. a. no limit, 1, 2, 8, 128, increases without bound
   b. limit = 0
   c. no limit, 2, 8, 128, increases without bound

Lesson 10 Square and Triangular Numbers

✎The sum of two consecutive triangular numbers equals a square number.
Use a formula to describe this relationship: $T_{n-1} + T_n = S_n$

Homework

2. $T_{10} = (10 \cdot 11)/2 = 55$

3. $T_{22} = (22 \cdot 23)/2 = 253$

4. a. $1^2 = 1$
   $11^2 = 121$
   $111^2 = 12321$
   $1111^2 = 1234321$

   b. 12345678987654321

   c. Many answers are possible, unless someone actually multiplies out the product.

   d. The exact value is 123456790120987654321. Most calculators show 1.23456790E20

5. a. 8

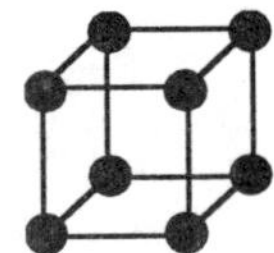

b. 27, One face has 9 dots on it. Imagine slicing the cube parallel to that face, through the remaining dots. You would need 2 more slices, each of which has 9 dots. (3slices)·(9 dots/slice) yields 27 dots.

c. n^3. To see this more clearly, imagine each cube cut vertically at each row of dots. Each slice has $n \cdot n$ dots. There are n of these slices. So the number of dots is $n \cdot n^2 = n^3$.

Lesson 11 Polygonal Number Patterns

Homework

2. No, $T_2 = 3$ is not a hexagonal number. There are many others.

Lesson 12 The Fibonacci Sequence

n-th Previous Generation	Number of Male Bees in the *n*-th Generation	*n*-th Previous Generation	Number of Male Bees in the *n*-th Generation
Current	1	6th	13
1st	1	7th	21
2nd	2	8th	34
3rd	3	9th	55
4th	5	10th	89
5th	8	11th	144

Homework

2. Looking at the female bees in each branch we get Fibonacci Numbers as well.

Lesson 13 The Golden Ration and the Fibonacci Numbers

✎ $F_1 = 1, F_2 = 2, F_3 = 2, F_4 = 3, F_5 = 5, F_6 = 8, F_7 = 13, F_8 = 21, F_9 = 34, F_{10} = 55$

Ratio of $\frac{F_{n+1}}{F_n}$	In decimals	Ratio of $\frac{F_{n+1}}{F_n}$	In Decimals
$\frac{F_2}{F_1} = \frac{1}{1}$	1	$\frac{F_7}{F_6} = \frac{13}{8}$	1.625
$\frac{F_3}{F_2} = \frac{2}{1}$	2	$\frac{F_8}{F_7} = \frac{21}{13}$	1.615385...
$\frac{F_4}{F_3} = \frac{3}{2}$	1.5	$\frac{F_9}{F_8} = \frac{34}{21}$	1.619048...
$\frac{F_5}{F_4} = \frac{5}{3}$	1.666666	$\frac{F_{10}}{F_9} = \frac{55}{34}$	1.617647...
$\frac{F_6}{F_5} = \frac{8}{5}$	1.6		

Lesson 14 Pascal's Triangle and Algebraic Patterns

✎ $(x + 1)^0 = 1$
$(x + 1)^1 = x + 1$
$(x + 1)^2 = x^2 + 2x + 1$
$(x + 1)^3 = x^3 + 3x^2 + 3x + 1$
$(x + 1)^4 = x^4 + 4x^3 + 6x^2 + 4x + 1$
$(x + 1)^5 = x^5 + 5x^4 + 10x^3 + 10x^2 + 5x + 1$
$(x + 1)^6 = x^6 + 6x^5 + 15x^4 + 20x^3 + 15x^2 + 6x + 1$
$(x + 1)^7 = x^7 + 7x^6 + 21x^7 + 35x^4 + 35x^3 + 21x^2 + 7x + 1$

Homework

1. $(x + 1)^8 = x^8 + 8x^7 + 28x^6 + 56x^5 + 70x^4 + 56x^3 + 28x^2 + 8x + 1$

2. $(x + 1)^9 = x^9 + 9x^8 + 36x^7 + 84x^6 + 126x^5 + 126x^4 + 84x^3 + 36x^2 + 9x + 1$

Lesson 15 Pattern Collections

none

Probability

Lesson 1 Introduction to Random Behavior

none

Lesson 2 Sample Spaces and Probability Models

Activity 1–Possible Outcomes

The sum of two dice
Outcomes:

	Die #1						
	Sum	1	2	3	4	5	6
Die #2	1	2	3	4	5	6	7
	2	3	4	5	6	7	8
	3	4	5	6	7	8	9
	4	5	6	7	8	9	10
	5	6	7	8	9	10	11
	6	7	8	9	10	11	12

Table 3 Sum of Two Dice

Sample space: {2, 3, 4, 5, 6, 7, 8, 9, 10, 11, 12 }

Activity 2–Constructing a Probability Model for a Two-Coin Toss

Outcome	Probability
2 heads	$\frac{1}{4} = 0.25$
1 head	$\frac{1}{2} = 0.5$
0 heads	$\frac{1}{4} = 0.25$

Table 6 Probability Model for Two-Coin Toss

Nitty Gritty: Conversion Among Percents, Fractions, and Decimals

Percent	Fraction	Decimal
25%	$\frac{1}{4}$	0.25
120%	$\frac{120}{100}=\frac{6}{5}$	1.2
266.6...%	$\frac{8}{3}$	2.6666...
20%	$\frac{1}{5}$	0.2
166.6...%	$\frac{5}{3}$	1.666...
83.3...%	$\frac{5}{6}$	0.8333...
100%	$\frac{100}{100}=1$	1.0

Table 8 Percentage, fraction and decimal forms

Homework

3. a. Assuming all numbers have occurred at least once, the chart indicates that the numbers 1 thorough 35 may be drawn.
   b. $\frac{5920}{5}=1184$ games
   c. For 11: $\frac{191}{1184}\approx 0.1613$. For 21: $\frac{147}{1184}\approx 0.1242$.
   d. $\frac{1}{35}\approx 0.0286$
   e. $5\cdot\frac{1}{35}=\frac{5}{1}\cdot\frac{1}{35}=\frac{5}{35}=\frac{1}{7}$
   f. The experimental result ranges from probability of 0.1242 to 0.1613, while the theoretical probability is 0.1429.
   g. Yes, if 1184 is a large number; No, if 1184 is not large enough.

Lesson 3 Empirical Probabilities

Activity 2–A Triple Coin Toss Thought Experiment

1. The sample space: {HHH, HHT, HTH, HTT, THH, THT, TTH, TTT }

2. 3 heads: HHH
   2 heads: HHT, HTH, THH
   1 head: HTT, THT, TTH
   0 heads: TTT

Nitty Gritty: Solving Linear Equations

1. $x + 12 = -31$
   $-12 \quad -12$
   $x = -43$

3. $-25.7 = x + (-34.9)$
   $+34.9 \quad +34.9$
   $x = 9.2$

5. $24x = 408$
   divide both sides by 24
   $\frac{24x}{24} = \frac{408}{24}$
   $x = 71$

7. $5195.7 = -63.7x$
   divide by -63.7
   $\frac{5195.7}{-63.7} = \frac{-63.7x}{-63.7}$
   $x \approx -81.565$

9. $\frac{3}{4}x = 84$
   multiply by 4/3
   $\frac{4}{3} \cdot \frac{3}{4}x = \frac{4}{3} \cdot 84$
   $x = 112$

11. $17x - 95 = 126$
    $+95 \quad +95$
    $17x = 221$
    divide by 17
    $\frac{17x}{17} = \frac{221}{17}$
    $x = 13$

13. $-12 + 6.5x = 40$
    $+12 \quad +12$
    $6.5x = 52$
    divide by 6.5
    $\frac{6.5x}{6.5} = \frac{52}{6.5}$
    $x = 8$

15. $14 = \frac{x}{12} + 5$
    $-5 \quad -5$
    $9 = \frac{x}{12}$
    multiply by 12
    $12 \cdot 9 = 12 \cdot \frac{x}{12}$
    $x = 108$

17. $12 - 6x + 4 = 5x - 5$
    $-6x + 16 = 5x - 5$
    $+6x \quad +6x$
    $16 = 11x - 5$
    $+5 \quad +5$
    $21 = 11x$
    divide by 11
    $\frac{11x}{11} = \frac{21}{11}$
    $x \approx 1.909$

19. $40 + 5(3x + 4) = 6(3 - x)$
    $15x + 60 = 18 - 6x$
    $+6x \quad +6x$
    $21x + 60 = 18$
    $-60 \quad -60$
    $21x = -42$
    divide by 21
    $\frac{21x}{21} = \frac{-42}{21}$
    $x = -2$

Lesson 4 Theoretical Probabilities

Homework

4. a. There are 30 ways to get a sum greater that 4. Thus the probability is $\frac{30}{36} = \frac{5}{6}$.

   b. There are 6 ways to get a sum less than 4. Thus the probability is $\frac{6}{36} = \frac{1}{6}$.

   c. There are 12 face cards in a deck. Thus the probability is $\frac{12}{52} = \frac{3}{13}$.

   d. There are 52 – 12 = 40 cards that are not face cards. Thus the probability is $\frac{40}{52} = \frac{10}{13}$.

   e. There are 51,011 winners out of 1,000,000 possible outcomes. Thus the probability is $\frac{51{,}011}{1{,}000{,}000} \approx 0.051$.

   f. There are 1,000,000 – 51,011 = 948,989 losing tickets. Thus the probability of losing is $\frac{948{,}989}{1{,}000{,}000} \approx 0.949$.

Lesson 5 The Basic Rules of Probability

Homework

5. a. Probability of not getting an ace = 1 – probability of getting an ace

   $= 1 - \frac{4}{52} = \frac{52-4}{52} = \frac{48}{52} = \frac{12}{13} \approx 0.923.$

   b. Probability of not getting 3 heads = 1 – probability of getting 3 heads

   $= 1 - \frac{1}{8} = \frac{8-1}{8} = \frac{7}{8} = 0.875.$

   c. Probability of not getting a 2 or a 12 = 1 – probability of getting a 2 or a 12

   $1 - \left(\frac{1}{36} + \frac{1}{36}\right) = 1 - \frac{2}{36} = \frac{36-2}{36} = \frac{34}{36} = \frac{17}{18} \approx 0.944$

Lesson 6 Expected Value of Equally Likely Outcomes

none

Lesson 7 Expected Value of Outcomes Not Equally Likely

Homework

2. Let a head have the value of 1 and a tail have the value of 0.
   EV = P(0 heads)·0 + P(1 head)·1 + P(2 heads)·2 + P(3 heads)·3
   $$EV = \frac{1}{8}\cdot 0+\frac{3}{8}\cdot 1+\frac{3}{8}\cdot 2+\frac{1}{8}\cdot 3$$
   $$= 0+\frac{3}{8}+\frac{6}{8}+\frac{3}{8}=1.5$$

4. Total area: $A=\pi r^2=\pi(18)^2\approx 1017.88\ in^2$
   Area 50: $A=\pi r^2=\pi 2^2\approx 12.57\ in^2$
   Area 40: $A=\pi 6^2-12.57\approx 100.53\ in^2$
   Area 30: $A=\pi 10^2-\pi 6^2\approx 201.06\ in^2$
   Area 20: $A=\pi 14^2-\pi 10^2\approx 301.59\ in^2$
   Area 10: $A=\pi 18^2-\pi 14^2\approx 402.13\ in^2$

   EV = P(50)·50 + P(40)·40 + P(30)·30 + P(20)·20 + P(10)·10
   $$=\frac{12.57}{1017.88}\cdot 50+\frac{100.53}{1017.88}\cdot 40+\frac{201.06}{1017.88}\cdot 30+\frac{301.59}{1017.88}\cdot 20+\frac{402.13}{1017.88}\cdot 10$$
   $$=\frac{628.5}{1017.88}+\frac{4021.2}{1017.88}+\frac{6031.8}{1017.88}+\frac{6031.8}{1017.88}+\frac{4021.3}{1017.88}=\frac{20734.6}{1017.88}\approx 20.37$$

Lesson 8 Expected Value vs. Outcome Probability

none

Lesson 9 Expected Value vs. Outcome Value

Activity 1–Powerball Lottery Game

Grand Prize = $20,000,000

$$EV=\frac{1}{54,979,155}\cdot 19,999,999+\frac{44}{54,979,155}\cdot 99,999+\frac{200}{54,979,155}\cdot 4999+\frac{8800}{54,979,155}\cdot 99$$
$$+\frac{343,200}{54,979,155}\cdot 4+\frac{98,800}{54,979,155}\cdot 4+\frac{456,950}{54,979,155}\cdot 1+\frac{658,008}{54,979,155}\cdot 0+\frac{53,405,352}{54,979,155}\cdot(-1)$$
$$=\frac{-24,137,247}{54,979,155}\approx -0.44$$

Grand Prize = $70,000,000

$$\text{EV} = \frac{1}{54{,}979{,}155}\cdot 69{,}999{,}999 + \frac{44}{54{,}979{,}155}\cdot 99{,}999 + \frac{200}{54{,}979{,}155}\cdot 4999 + \frac{8800}{54{,}979{,}155}\cdot 99$$
$$+\frac{343{,}200}{54{,}979{,}155}\cdot 4 + \frac{98{,}800}{54{,}979{,}155}\cdot 4 + \frac{456{,}950}{54{,}979{,}155}\cdot 1 + \frac{658{,}008}{54{,}979{,}155}\cdot 0 + \frac{53{,}405{,}352}{54{,}979{,}155}\cdot(-1)$$
$$= \frac{25{,}862{,}753}{54{,}979{,}155} \approx 0.47$$

$$\text{EV} = \frac{1}{54{,}979{,}155}\cdot x + \frac{44}{54{,}979{,}155}\cdot 99{,}999 + \frac{200}{54{,}979{,}155}\cdot 4999 + \frac{8800}{54{,}979{,}155}\cdot 99$$
$$+\frac{343{,}200}{54{,}979{,}155}\cdot 4 + \frac{98{,}800}{54{,}979{,}155}\cdot 4 + \frac{456{,}950}{54{,}979{,}155}\cdot 1 + \frac{658{,}008}{54{,}979{,}155}\cdot 0 + \frac{53{,}405{,}352}{54{,}979{,}155}\cdot(-1) = 0$$

$$\frac{x}{54{,}979{,}153} - \frac{44{,}137{,}246}{54{,}979{,}153} = 0$$

$x = 44{,}137{,}246$

So when the Grand Prize is $44,137,247 the expected value is 0.

Lesson 10 Final Paper

none

Lesson 11 Extensions Conditional Probability

Homework

2.

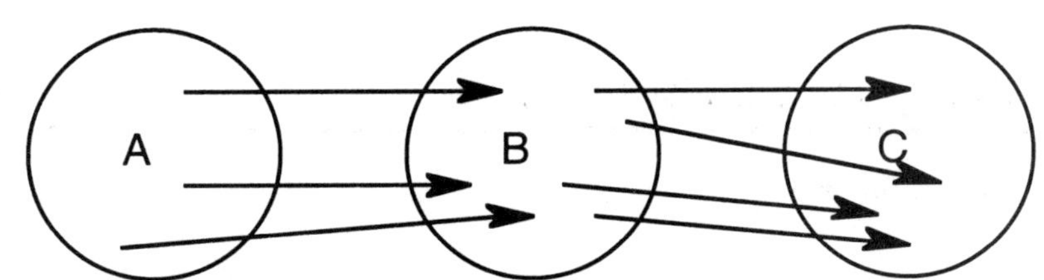

a. 4 b. 4 c. 4 d. 12

e. $\frac{1}{12} = 0.083\overline{3}$ f. $\frac{1}{12}$

g. Fundamental Counting Principle. In both cases, there are two parts to the problem. For the first part the probability is 1 out of 3 or 1/3. For the second part, the probability is 1 out of 4 or 1/4.

6.

		FIRST SHOT P(Hit) = H	P(Miss) = M
SECOND SHOT	P(Hit) = H	$H \cdot H = H^2$	$M \cdot H = HM$
	P(Miss) = M	$H \cdot M = HM$	$M \cdot M = M^2$

$H^2 + 2HM + M^2 = 1$

Representations of Data

Lesson 1 Fishbone Diagrams

none

Lesson 2 Charts – An Analysis Using Wage Charts

Activity 2–Organizing Numerical Data

Sample Wage Chart 1, Arranged Numerically by Associate Degree

	Associate	Bachelor	Master	Doctoral
Systems Analysis	32333	43333	54444	65444
Actuarial	29251	37251	45251	53251
Materials Engineering	27170	37700	48230	58230
Mechanical Engineering	26970	37500	48030	58030
Information Sciences	25666	36666	47777	58777
Management Information Systems	24799	34799	44799	54799
Computer Science	24784	35784	46895	57895
Systems Engineering	23470	34000	44530	54530
Engineering Technology	23270	33800	44330	54330
Distribution Management	22305	32305	42305	52305
Accounting	21511	31511	41511	51511
Safety Engineering	21470	32000	42530	52530
Mathematics	21316	29316	37316	45316
Economics & Finance	20610	30610	40610	50610
Business Administration	18961	28961	38961	48961
Architectural & Environmental Design	17267	25267	33267	41267

Homework

2. a.

Subject Area for Degree	Starting Salary per Years of Study			
	Associate	Bachelor	Master	Doctoral
BUSINESS				
Accounting	10756	7878	6919	5723
Business Administration	9481	7240	6494	5440
Distribution Management	11153	8076	7051	5812
Economics & Finance	10305	7653	6768	5623
Management Information Systems	12400	8700	7467	6089
ENGINEERING				
Engineering Technology	11635	8450	7388	6037
Materials Engineering	13585	9425	8038	6470
Mechanical Engineering	13485	9375	8005	6448
Safety Engineering	10735	8000	7088	5837
Systems Engineering	11735	8500	7422	6059
COMPUTER SCIENCES				
Computer Science	12392	8946	7816	6433
Information Sciences	12833	9167	7963	6531
Systems Analysis	16167	10833	9074	7272
SCIENCES				
Actuarial	14626	9313	7542	5917
Architectural & Environmental Design	8634	6317	5545	4585
Mathematics	10658	7329	6219	5035

b. Shelly should pursue an Associate degree in Computer Science System Analysis.

Nitty Gritty: Arithmetic Sequences and Their Sums

1. $a_5 = 2 + (5 - 1) \cdot 6$
$= 2 + 4 \cdot 6$
$= 26$

3. $a_1 = -3, d = -6 - (-3) = -3$
$a_{10} = -3 + (10 - 1) \cdot (-3)$
$= -30$

5. $3 + 6 + 9 + 12 + 15 = 45$

$$S_5 = \frac{5}{2}[2 \cdot 3 + (5-1) \cdot 3]$$
$$= \frac{5}{2}(6 + 4 \cdot 3)$$
$$= \frac{5}{2}(18)$$
$$= 45$$

7. $a_1 = \frac{1}{3}, d = \frac{1}{3}$

$$S_7 = \frac{7}{2}\left(\frac{2}{3} + 6 \cdot \frac{1}{3}\right)$$
$$= \frac{7}{2} \cdot \frac{8}{3}$$
$$= \frac{28}{3}$$

9. $6 - 3 = 3,\ 12 - 6 = 6$
This is not an arithmetic sequence. We can still find the sum of the first five terms.
$S_5 = 3 + 6 + 12 + 24 + 48$
$= 93$

11. $-0.1 - (-0.2) = 0.1$
$0 - (-0.1) = 0.1$, etc
$a_1 = -0.2,\ d = 0.1$
$S_{10} = \frac{10}{2}\left[2\cdot(-0.2) + 9\cdot(0.1)\right]$
$= 5\cdot(-0.4 + 0.9)$
$= 5(0.5)$
$= 2.5$

Nitty Gritty: Geometric Sequences and their Sums

1. $a_5 = 2\cdot 3^{5-1}$
$= 2\cdot 3^4$
$= 162$

3. $a_1 = 3,\ r = \frac{6}{3} = 2$
$a_8 = 3\cdot 2^7$
$= 384$

5. $a_1 = 10,\ r = \frac{50}{10} = 5$
$S_6 = \frac{10(1-5^6)}{1-5}$
$= 39{,}060$

7. $a_1 = \frac{1}{3},\ r = \frac{2/3}{1/3} = 2$
$S_7 = \frac{\frac{1}{3}(1-2^7)}{1-2}$
$= \frac{127}{3}$

9. $\frac{7/4}{1/4} = 7,\ \frac{13/4}{7/4} = \frac{13}{7}$
This is not a geometric sequence.

11. $\frac{0.1}{0}$ is undefined, $\frac{0.2}{0.1} = 2,\ \frac{0.3}{0.2} = 1.5$
This is not a geometric sequence.

Lesson 3 Graphical Representations

Activity 1–Extracting Information from a Matrix

1. $a_{1,4}$: guppy
$a_{3,4}$: marlin
$a_{3,1}$: lion
$a_{3,2}$: eagle
$a_{3,3}$: crocodile

3. Give the matrix address for each of the following:
$SA_{8,4}$ \$112,160
$AED_{1,1}$ \$17,267
$SA_{8,1}$ \$39,766

4. To determine the average first year's salary, we must add all the first year salaries for employees with a Doctorate and divide by 16. This gives $53,611.63.

Activity 2–Whiskergraph and Histogram

1. Specific animals and their approximate sizes are included in matrix A but not the Whiskergraph.

3. Reptiles vary in size the most since the Whiskergraph is the longest.

9. A person with a Master's and 8 years service will make more than a person with a doctorate and only 1 to 3 years service. A person with a Master's and 7 years service will make more than a person with a doctorate and only 1 to 2 years service. A person with a Master's and 5 to 6 years service will make more than a person with a doctorate and only 1 year of service.

13. The average salary of accountants with between 1 and 8 years service and an Associate degree is $23910.38.

14. To answer the previous three questions use a matrix. The specific salary values are not easily extracted from the histogram.

Activity 4–Analyzing Specific Information from a Matrix

1.

ACC	BA	DM	EF	MIS
645	569	669	618	744

4.

ACC	BA	DM	EF	MIS
1.03	1.03	1.03	1.03	1.03

8. The total income is $378,323, if you total the Bachelor column for the first 8 years then add $41466 \cdot 1.04$ and $41466 \cdot 1.04^2$ and so on.

$$31511 + 32771 + 34082 + 35446 + 36863 + 38338 + 39871 + 41466 + 43125 + 44850 = 378,323$$

9. The total income is $547,148.

Homework

1. a.

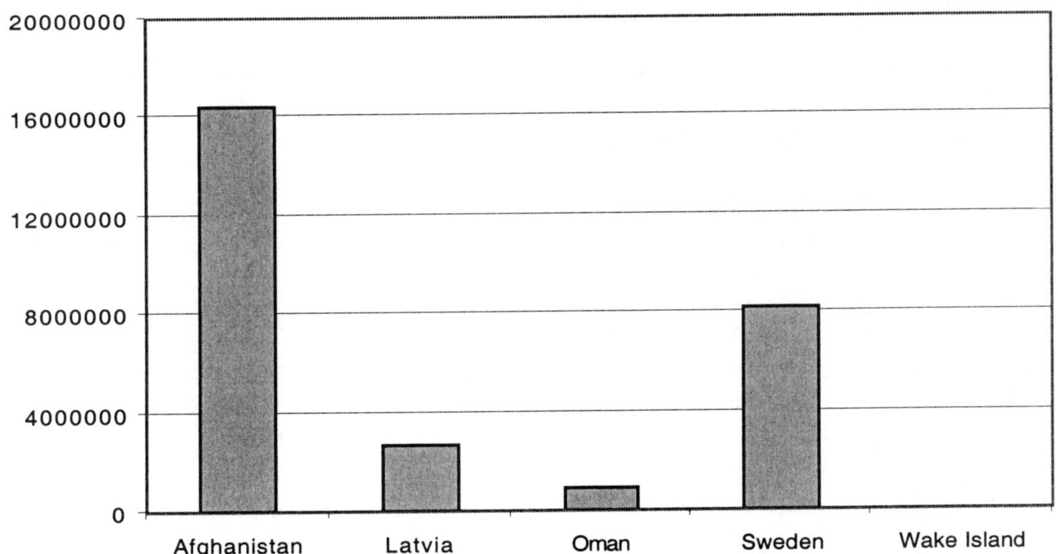

b. The information best represented by the matrix is specific population values.

c. The information that is best represented by this graphical representation are the relative differences between populations.

2. a.

Market Highlights

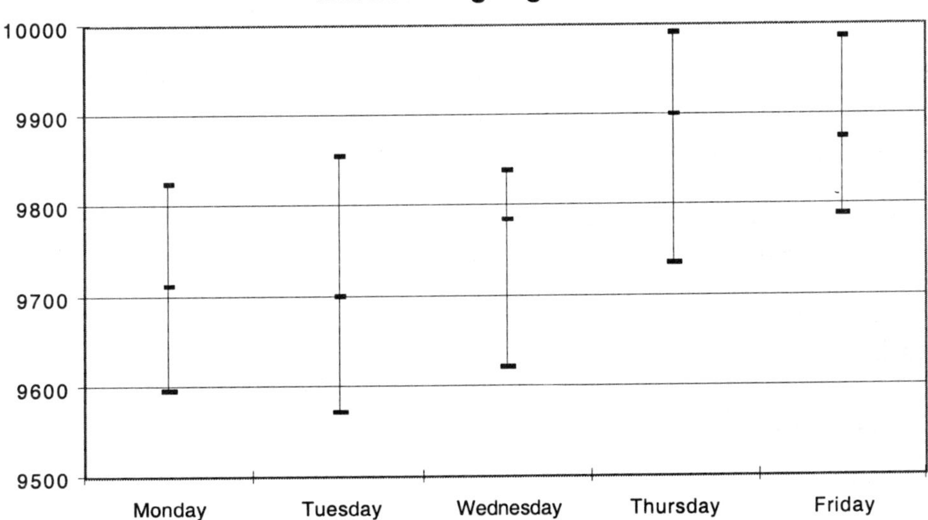

b. The information best represented by the matrix is specific market values (e.g. the low on Thursday was 9,735).

c. The information best represented by your graphical representation is the range of values for each day (e.g. on which day the market varied the most) and general trends between days.

3. a. Assuming that they all start at level one, you will pay \$651,257 over the next four years.

Assoc: $22305\frac{1.03^4-1}{0.03}=93316$

Bachelors: $32305\frac{1.04^4-1}{0.04}=137182$

Masters: $42305\frac{1.06^4-1}{0.06}=185068$

Ph.D.: $52305\frac{1.08^4-1}{0.08}=235692$

b. The arithmetic sequence formula could be used to sum the salaries for all four degree levels for the first year then the remaining salaries for the second, third and fourth years could be added. The arithmetic sequence formula can not be used to sum the salaries for the second, third or fourth year because the differences between the salaries for each year are not the same. The geometric sequence formula could be used to sum the salaries for the first four years for each degree level, then the total salaries for each degree level could be summed.

Lesson 4 Bar Charts, Line Graphs and Matrices

Activity 1–Procured Parts Lead-Time

1. The amount of time spent in inventory (lead-time) seems to be decreasing.

2. The stacked bar chart is visual; for most people it is easier to read and understand.

3. For 3/31/97, $(22\%)(10)+(33\%)(30)+(44\%)(85)\approx 49.5$ days.
For 5/31/97, $(22\%)(10)+(44\%)(30)+(33\%)(85)\approx 43.5$ days.
Goal: $(56\%)(10)+(44\%)(30)\approx 18.8$ days.

9. It may be tempting for the students to conclude that the number of parts in inventory is decreasing. This cannot be concluded because the chart only shows changes in lead-time. It contains no information about the specific quantities of parts in inventory.

Activity 2–Sales Backlog

1. Total Sales = Backlog + Future Backlog

2. Future Backlog = 1st 30 + (31+days)

4. Either answer is fine, so long as there is a supporting explanation. With a line graph it is easy to see when backlog line is at lowest point and when future backlog line is at highest point. When the line graph was too cluttered, using a matrix can help. You could also just look for smallest future backlog value by scanning a column of numbers.

6. Use a matrix because it is difficult to read exact values from a line graph.

8. Use a line graph because it is easy to see trends (increasing, decreasing).

Activity 3–Manufacturing Cycle Times

1. The best cycle time was for the Model 50 and the worst cycle time is for the Model 75.

2. Use a bar graph because it is easy and quick to detect the tallest and shortest bar.

4. Use a matrix because it is difficult to read specific values from a bar chart.

6. The manufacturing cycle time of a product is found by dividing total hours by total end-to-end lead time. This fraction (ratio) can be made larger by increasing the numerator (total hours) or by decreasing the denominator (end to end lead time).

Homework

3. a. The top line, total sales, is the sum of the bottom two lines. The backlog represents the product that the customer is waiting for. The total future line represents the product that is scheduled to be delivered in the future.

   b. Pro: This graph breaks down specific categories of sales and shows how they relate to each other. Con: The range of times for future deliveries is not shown by this graph.

5. a. The graph shows the manufacturing cycle time percent for 4 different models of pop dispensing machines. Manufacturing cycle time percent is defined as the actual time it takes to build the product divided by the end to end lead time.

   b. Pro: Different models are compared. Pro: Goals can be established based on comparative values for different models. Con: Percentage does not give an indication of actual hours end to end lead time.

Lesson 5 Coded Maps – Computer Parts Suppliers

Activity 1–Coded Map Examples

- Elevation maps
- Rainfall maps
- Image maps on the world-wide-web
- Calendar with phases of the moon
- Crochet patterns
- A treasure map

Activity 2–Creating a Coded Map

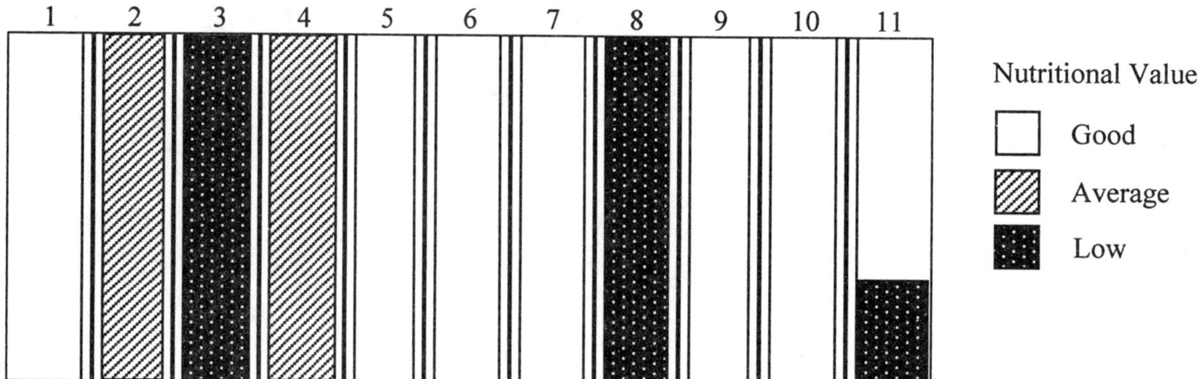

1. The information that is emphasized in the coded map is the nutritional value of grocery items found in areas of the store.

2. The information that is emphasized in the store directory is the specific location of specific grocery items.

5. A store directory chart provides the more information because all the information in the coded map can be retrieved from the store directory. In addition, the directory tells you the exact location of specific foods.

Activity 3–Extracting Information from a Coded Map

2. The companies in question 1 are not certified.

5. Answers will vary. The coded map in Figure 13 emphasizes delivery time.

7. Yes, TU is certified. Figure 13 was used because the list of all suppliers appear close together and are easy to scan quickly.

Homework

2. a. The total population is 725,000.
   b. The total number of store is 20.
   c. The average number of people per store is 36,250.
   d. Zone A has 225,000 people; Zone B 100,000; Zone C 300,000; and Zone D has 100,000.
   e. Zone A has 7 stores, Zone B has 3, Zone C has 4, and Zone D has 6 stores.
   f. The average number of people per store in Zone A is 32,143; in Zone B 33,333; Zone C has 75,000 and Zone D has 16,667.
   g. Probably not. If you look at the map, you will see that the majority of Zone C's population is concentrated near the border of the other three zones and is located very close to 8 retail stores. This gives that population area an average of 31,250 people per store which is below the average. There is a 100,000 population area in Zone B that is not very well served. That might be a better location for a new store. Of course, there are also other factors involved beside population. For example: taxes, zoning laws, labor availability, average income in area, environment, availability of parking.

h. Taxes, zoning laws, labor availability, average income in area, environment, availability of parking.

Lesson 6 Decision Trees

Activity 1–Use a Decision Tree

1. The vendor CP delivers in the least time according to the coded map. The decision tree only indicates that TU and CP deliver in 1-2 days. It does not indicate which or whether, TU or CP, delivers in the least time.

2. AD and HJ both have 8 parts below average price. Neither representation really appears to be better than the other. The coded map requires counting lightly shaded areas for each supplier. The decision tree requires counting the number of entries in the 'Below Average' branch for each vendor.

6. NP will deliver in exactly 3 days. The decision tree contains specific delivery times, the coded map does not.

Homework

3. a. Characteristics of a decision tree:
   Answers questions with two or more choices
   Linear format with no feedback/looping
   A problem solving technique
   Uses filters to find the answer to a problem

   Characteristics of a flow chart:
   Answers question with two choices, usually yes/no
   Can be linear or circular, usually has feedback/looping
   A problem solving technique
   Uses filters to find the answer to a problem

Lesson 7 Process Control Charts – Potato Chip Quality Analysis

Activity 4–Analysis

1. Two obvious situations in which the limits would be unreasonable are: too lax (the calculated limits may allow a bag to be full of crushed chips and still be considered quality chips) or too stringent (the calculated limits may insist that a bag have very few crushed chips to be considered quality chips).

2. If the sample size was too small and just happened to be a bag of a perfect or crushed bag of chips. If the bags of chips had been badly roughed up before being used in these activities. It is important to get a representative sample. It could also occur if the average quality of the chips was, and always is, bad.

3. A high mean should indicate good quality, but the high mean may have come about from a lot of bags with high whole chips per ounce (WC/O) measures and a few bags with extremely low WC/O measures. If you are the customer who gets the bag with extremely low WC/O you may not think the product shows high quality.

   A low variance indicates that each bag will have about the same WC/O as any other bag. This would be good if the WC/O was high, it would be bad if the WC/O was low.

4. a. A high mean would tend to indicate high quality.
   b. A low variance would tend to indicate high quality.
   c. A high mean and low variance simultaneously would indicate the highest quality.

Homework

2.

data	data – mean	square of (data – mean)
12	12.6	158.76
15	15.6	243.36
–13	–12.4	153.76
–19	–18.4	338.56
23	23.6	556.96
–21	–20.4	416.16
15	15.6	243.36
–16	–15.4	237.16
18	18.6	345.96
–20	20.6	376.36
Total = –6		Total = 3074
Mean = –0.6		Variance = 307.4
		Sigma = 17.5
upper control limit = 51.9		
lower control limit = –53.1		

data	data – mean	square of (data – mean)
25	–3.6	12.96
32	3.4	11.56
31	2.4	5.76
27	–1.6	2.56
28	–0.6	0.36
22	–6.6	43.56
32	3.4	11.56
33	4.4	19.36
19	–9.6	92.16
37	8.4	70.56
Total = 286		Total = 270.4
Mean = 28.6		Variance = 27.04
		Sigma = 5.2
upper control limit = 44.2		
lower control limit = 13		

Lesson 8 Representations as Aids in Reporting

✎ $550 \cdot 0.018 \approx 10$

Homework

1.

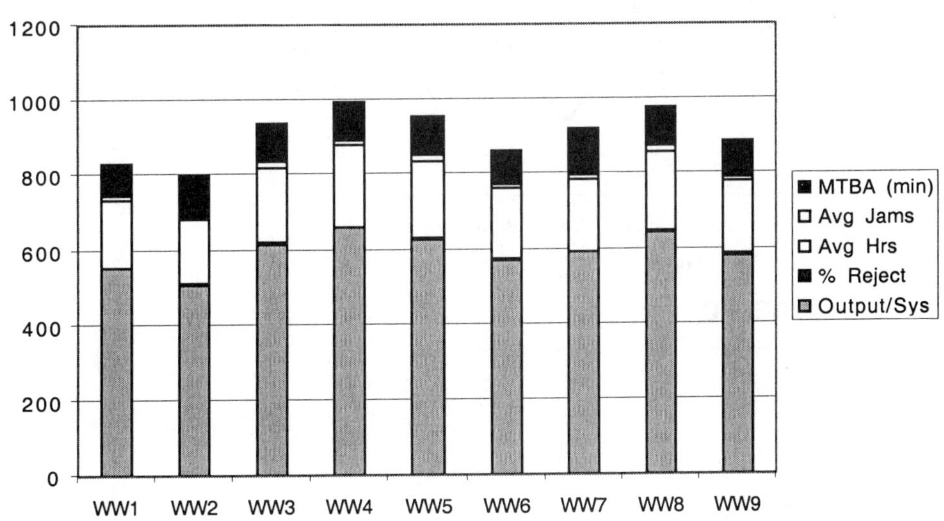

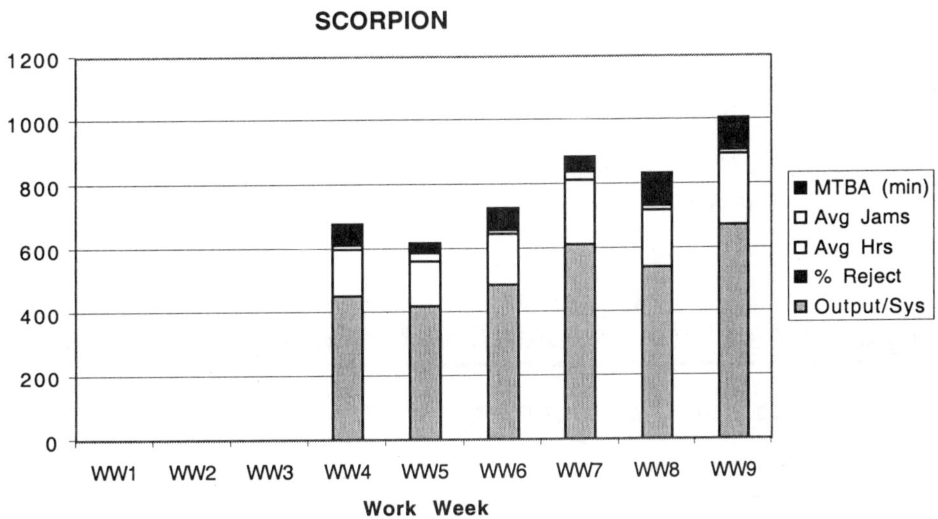

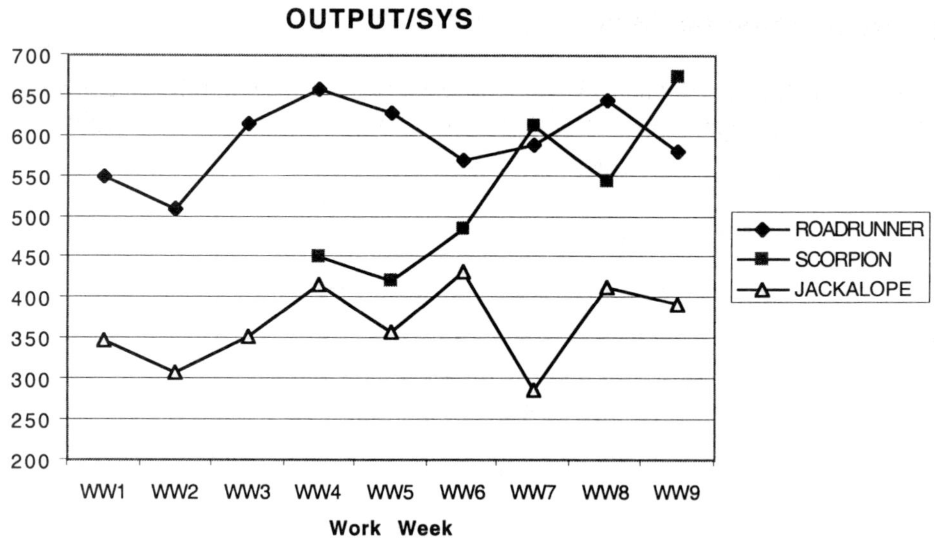

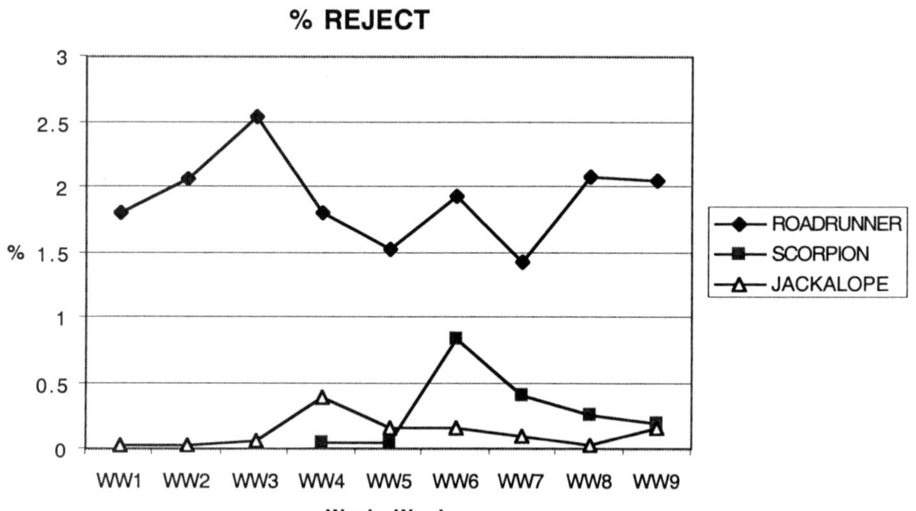

Lesson 9 Using Graphical Representation for Data Comparison

Activity 1–Combining Data from Two Sources

1., 2.

Output/Sys vs. Safety Trends									
	WW1	WW2	WW3	WW4	WW5	WW6	WW7	WW8	WW9
RR Output/Sys	550	509	615	658	628	571	589	645	582
JA Output/Sys	347	308	353	418	358	433	286	413	392
SC Output/Sys				452	423	485	613	543	674
Days Lost	2	4	2	1	2	3	2	1	2

Activity 2–Graph the Combined Data

By creating a scatterplot of the totals for Output/Sys versus Days Lost, we see that the data is widely scattered indicating that there is no apparent relationship. This is easier to see in the scatterplot.

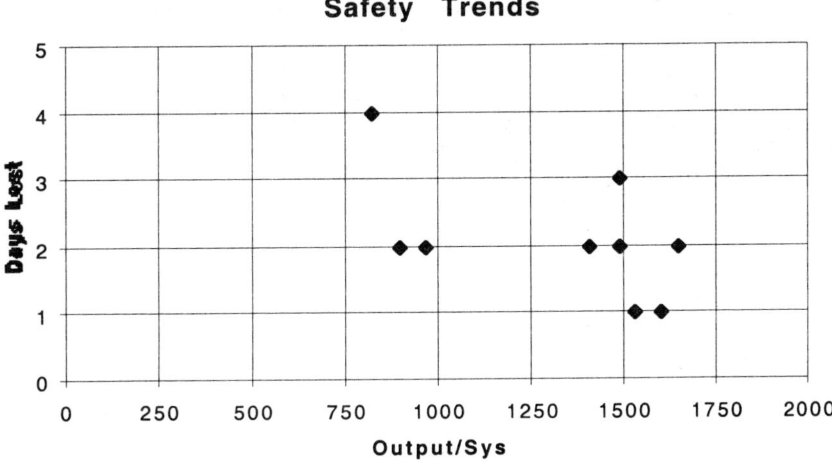

Wrap Up

Answers will vary, but should include (1) choose type of graph, (2) choose to data to be recorded on the vertical and horizontal axes, (3) pick scale and (4) graph.

Homework

2. Example: Reference: "Camelback Corridor builds on its prestige". Arizona Republic, section E, May 29, 1997.

YEAR	METRO VACANCY %	CAMELBACK VACANCY %
1990	26	30
1991	26	30
1992	25	25
1993	20	23
1994	16	12
1995	14	11
1996	12	8
1997	9.6	4.6

Explanation:
The article is a little misleading in that it gives incomplete information:

- The article does not give information as to the total number of business sampled vs. the total number of businesses in the area.
- It is unknown how much new office space was built in the two areas during the timeframe given.
- No explanation was given as to why Metro Vacancy was 3-4% less during 1990-1993, and then Camelback Vacancy was 3-5% less during 1994-1997.

Right Triangle Trigonometry

Lesson 1 Angles and Their Triangle Ratios

Homework

2. a. 53°
   $53° + 360° = 413°$
   $53° + 2 \cdot 360° = 773°$
   $53° + 3 \cdot 360° = 1133°$

   b. 0°
   $0° + 360° = 360°$
   $0° + 2 \cdot 360° = 720°$
   $0° + 3 \cdot 360° = 1080°$

   c. 330°
   $330° + 360° = 690°$
   $330° + 2 \cdot 360° = 1050°$
   $330° + 3 \cdot 360° = 1410°$

   d. −125°
   $-125° - 360° = -485°$
   $-125° - 2 \cdot 360° = -845°$
   $-125° - 3 \cdot 360° = -1205°$

4. $\frac{\sin(65°)}{\cos(65°)} \approx \frac{0.93}{0.40} \approx 2.325$ and $\tan(65°) \approx 2.33$

   The two values, $\frac{\sin(65°)}{\cos(65°)}$ and $\tan(65°)$, are close enough to suggest that they should be equal.

   Note: the values used for sin(65°), cos(65°), and tan(65°) come from the answers provided for Activity 1, and may differ slightly from your values.

angle	sin(angle)	cos(angle)	$\frac{\sin(\text{angle})}{\cos(\text{angle})}$	tan(angle)
33°	0.5446	0.8387	0.6494	0.6494
47°	0.7314	0.6820	1.0724	1.0724
5°	0.0872	0.9962	0.0875	0.0875
90°	1.0000	0.0000	undefined	undefined
0°	0.0000	1.0000	0.0000	0.0000

$$\tan(\text{angle}) = \frac{\sin(\text{angle})}{\cos(\text{angle})}$$

Nitty Gritty–The Pythagorean Theorem

1. 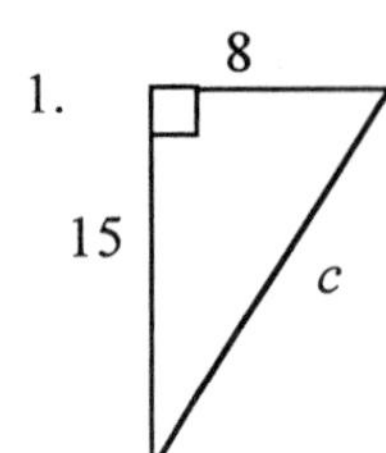

$$a^2 + b^2 = c^2$$
$$8^2 + 15^2 = c^2$$
$$64 + 225 = c^2$$
$$289 = c^2$$
$$\pm\sqrt{289} = c, \text{ choose } +$$
$$17 = c$$

3.

40

41

c

$$a^2 + b^2 = c^2$$
$$40^2 + c^2 = 41^2$$
$$1600 + c^2 = 1681$$
$$c^2 = 81$$
$$c = \pm\sqrt{81}\text{, choose } +$$
$$c = 9$$

5.

12

7

a

$$a^2 + b^2 = c^2$$
$$a^2 + 7^2 = 12^2$$
$$a^2 + 49 = 144$$
$$a^2 = 95$$
$$a = \pm\sqrt{95}\text{, choose } +$$
$$a \approx 9.7468$$

7.

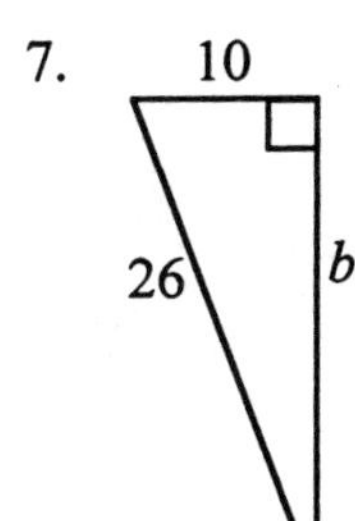

$$a^2 + b^2 = c^2$$
$$10^2 + b^2 = 26^2$$
$$100 + b^2 = 676$$
$$b^2 = 576$$
$$b = \pm\sqrt{576}\text{, choose } +$$
$$b = 24$$

9. a. Draw and label a picture:

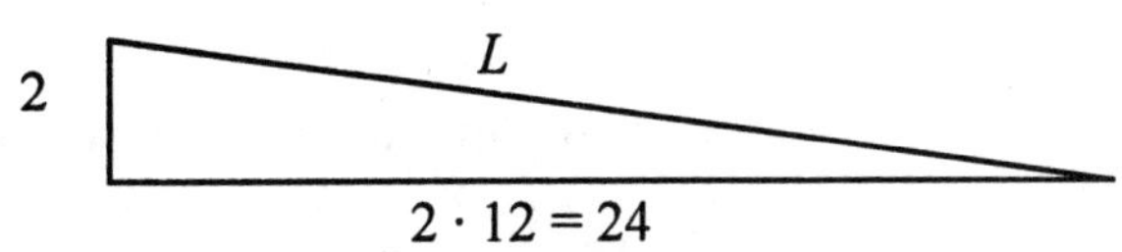

b. Use the Pythagorean Theorem to solve for the length of the ramp, L.

$$a^2 + b^2 = L^2$$
$$2^2 + 24^2 = L^2$$
$$4 + 576 = L^2$$
$$580 = L$$
$$\pm\sqrt{580} = L\text{, choose } +$$
$$24.083 \approx L$$

c. The length of the ramp will be 24.083 feet.

Lesson 2 Solving Right Triangles

Homework

3. Using the results of 1. and 2., complete the table using exact values:

$\sin(\angle F) = \sin(30°) = \frac{1}{2}$ $\cos(\angle F) = \cos(30°) = \frac{\sqrt{3}}{2}$ $\tan(\angle F) = \tan(30°) = \frac{1}{\sqrt{3}}$

$\sin(\angle A) = \sin(45°) = \frac{1}{\sqrt{2}}$ $\cos(\angle A) = \cos(45°) = \frac{1}{\sqrt{2}}$ $\tan(\angle A) = \tan(45°) = \frac{1}{1} = 1$

$\sin(\angle G) = \sin(60°) = \frac{\sqrt{3}}{2}$ $\cos(\angle G) = \cos(60°) = \frac{1}{2}$ $\tan(\angle G) = \tan(60°) = \frac{\sqrt{3}}{1}$

$\sin(\angle B) = \sin(90°) = \frac{\sqrt{2}}{\sqrt{2}} = 1$ $\cos(\angle B) = \cos(90°) = 0$ $\tan(\angle B) = \tan(90°) = \frac{1}{0}$, undefined

5. In order for the pictured ramp to meet the ADA guidelines, the run would need to be at least 48 feet long (4 · 12).

$\tan(10°) = \frac{4}{\text{run}}$

4 ft 10° run

$\text{run} = \frac{4}{\tan(10°)} \approx 22.6851$ feet. The ramp does not meet the ADA guidelines.

Lesson 3 The Unit Circle

Homework

2. a. sin (360°) is the same as sin (0°). Since sin(0°) = 0, sin (360°) = 0.

b. $\tan(225°) = \frac{\sin(225°)}{\cos(225°)} = \frac{-\frac{1}{\sqrt{2}}}{-\frac{1}{\sqrt{2}}} = 1$

c. $\cos(-135°) = \cos(225°) = -\frac{1}{\sqrt{2}}$

d. $\tan(135°) = \frac{\sin(135°)}{\cos(135°)} = \frac{\frac{1}{\sqrt{2}}}{-\frac{1}{\sqrt{2}}} = -1$

e. $\cos(180°) = -1$

(–1,0)

f. $\tan(270°) = \frac{\sin(270°)}{\cos(270°)} = \frac{-1}{0}$ which is undefined

g. $\tan(-270°) = \tan(90°) = \dfrac{\sin(90°)}{\cos(90°)} = \dfrac{1}{0}$ which is undefined

h. $\sin(0°) = 0$

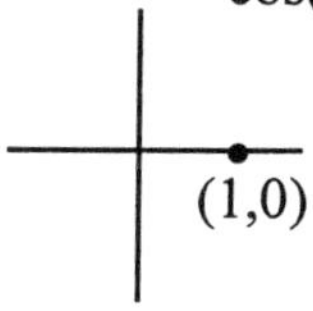

(1,0)

i. $\cos(360°) = \cos(0°) = 1$

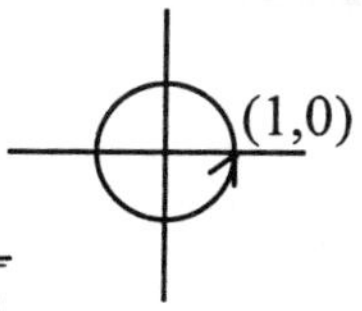

(1,0)

j. $\cos(315°) = \cos(45°) = \dfrac{1}{\sqrt{2}}$

k. $\sin(-210°) = \sin(150°) = \sin(30°) = \dfrac{1}{2}$

l. $\tan(900°) = \tan(180°) = \dfrac{\sin(180°)}{\cos(180°)} = \dfrac{0}{1} = 0$

6. a. $[\sin(30°)]^2 + [\cos(30°)]^2 = [\frac{1}{2}]^2 + [\frac{\sqrt{3}}{2}]^2$
$= \frac{1}{4} + \frac{3}{4}$
$= 1$

b. $[\sin(60°)]^2 + [\cos(60°)]^2 = [\frac{\sqrt{3}}{2}]^2 + [\frac{1}{2}]^2$
$= \frac{3}{4} + \frac{1}{4}$
$= 1$

c. $[\sin(90°)]^2 + [\cos(90°)]^2 = [1]^2 + [0]^2$
$= 1$

d. $[\sin(A)]^2 + [\cos(A)]^2 =$ appears to always be equal to 1, regardless of the value of A.

Lesson 4 Applications and Extensions

Activity 1–Which Quadrant?

1. The terminal side is in the third quadrant, in which x's are negative and the y's are negative. Accordingly, the tangent is positive and the cosine is negative.
2. Cosines are positive in the first and fourth quadrants, while tangents are negative in the second and fourth quadrants. Accordingly, cosines are positive and tangents are negative in the fourth quadrant.
3. Sines are negative in the third and fourth quadrants, while tangents are positive in the first and third quadrants. Accordingly, sines are negative and tangents are positive in the third quadrant.
4. Sines are positive in the first and second quadrants, while tangents are positive in the first and third quadrants. Accordingly, the sine is positive (0.2) and the tangent is positive in the first quadrant.

5. Sines are positive in the first and second quadrants, while tangents are negative in the second and fourth quadrants. Accordingly, the sine is positive and the tangent negative (–0.5) in the second quadrant.

Lesson 5 Self-Test and Project Planning

Basic Trigonometry Skills Self-Test:

1. The sum of the measures of the angles in a triangle is 180°.
2. The Pythagorean Identity is $[\sin(A)]^2 + [\cos(A)]^2 = 1$
3. $\sin(Q) = \frac{o}{m}$, $\cos(P) = \frac{o}{m}$, and $\tan(Q) = \frac{o}{n}$
4.

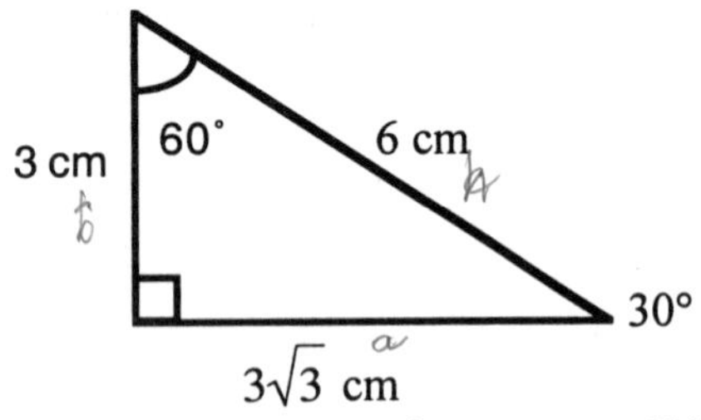

Find the unknown angle measure: $60° + 90 + x° = 180°$; so the unknown angle is 30°.
Find the length of the hypotenuse:

$\cos(60°) = \frac{1}{2}$ known cosine value of 60° angle

$\cos(60°) = \frac{\text{adjacent}}{\text{hypotenuse}} = \frac{3}{h}$ trigonometric ratio

$\frac{1}{2} = \frac{3}{h}$ set cosine values equal

$h = 6$ solve to find the hypotenuse

Find the length of the unknown side, a.

$a^2 + b^2 = c^2$ Pythagorean Theorem

$a^2 + 3^2 = 6^2$ substitute known values for b and c

$a^2 = 27$ square and subtract 9 from both sides

$a = \sqrt{27}$ take the square root of both sides

$a = 3\sqrt{3}$ simplify $\sqrt{27}$.

5. Maximum for $\cos(A)$: +1
Minimum for $\cos(A)$: –1

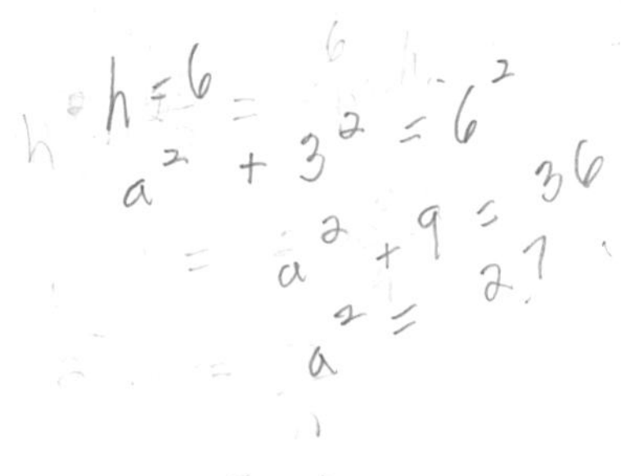

6. a. Draw angle F so that: sin (F) is positive and cos (F) is negative.
Draw any angle with terminal side in the second quadrant.

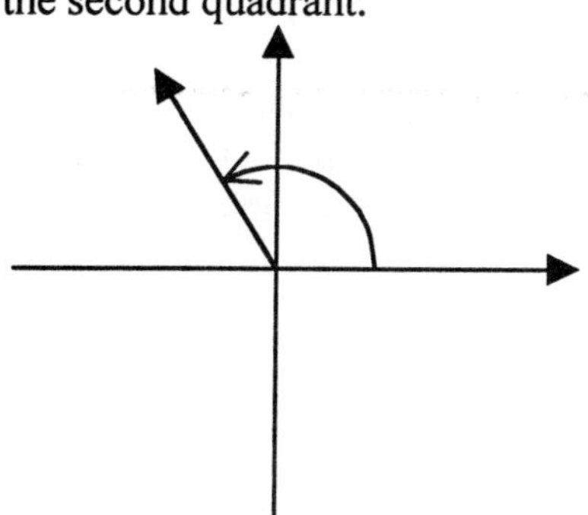

b. Draw angle H so that: cos (H) is positive and tan (H) is negative.
Draw any angle with terminal side in the fourth quadrant.

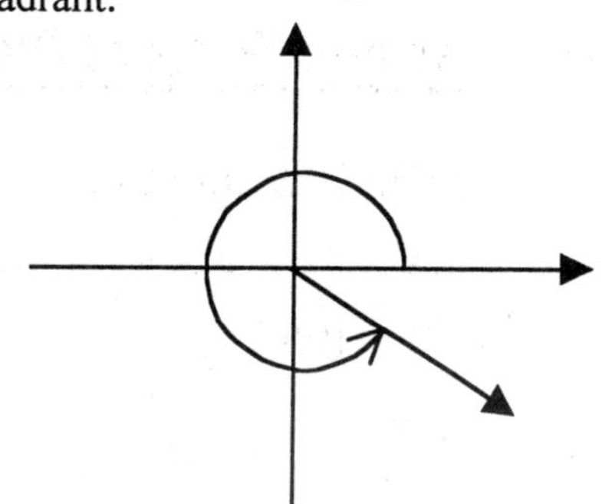

7. The problem states that the roof has a slope of 1/3. Recall that the numerator indicates the *rise* and the denominator indicates the *run*. A slope of 1/3 means that for every 1 foot of rise there are 3 feet of run. The roof is 5 feet high, or has a rise of 5 feet. Accordingly, it must have a run of 5 · 3 = 15 feet. Note that the 15-foot run represents half the span of the roof. So, the span is 30 feet long.

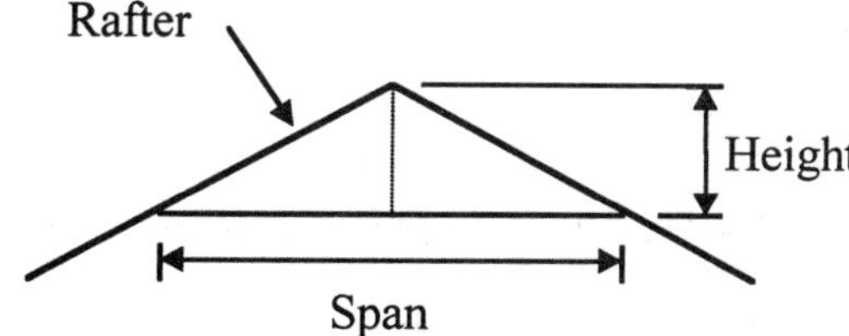

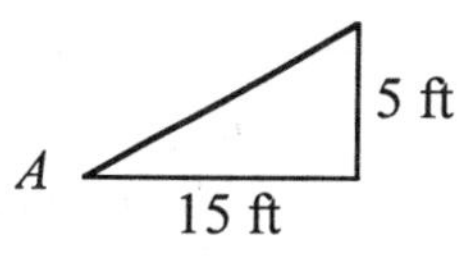

To find the angle of the rafters, consider a right triangle representing one-half of the roof. One side is 5 ft and the other side is 15 ft. The 5-ft side is opposite the rafter angle, A, and the 15-foot side is adjacent to the rafter angle. The tangent of A can be used to find the rafter angle, since it involves the opposite and adjacent sides.

$$\tan(A) = \frac{\text{opposite}}{\text{adjacent}} = \frac{5}{15} = \frac{1}{3}$$

$\angle A = \tan^{-1}(1/3) = 18.43495°$.

To find the length of the rafters, the Pythagorean Theorem can be used. Notice that the rafters represent the hypotenuse of a right triangle.

$a^2 + b^2 = c^2$	Pythagorean Theorem
$5^2 + 15^2 = c^2$	substitute known values for a and b
$25 + 225 = c^2$	square
$250 = c^2$	add
$c = \sqrt{250} \approx 15.81139$ feet	take the square root of both sides

8. The length of the ship, 1000 ft, is opposite the 4° angle, and the distance from the shore, x, is adjacent to the 4° angle. Accordingly, a trigonometric ratio involving the opposite and the adjacent sides could be used to find the distance from the shore. the tangent involves the opposite and adjacent sides.

4°

x

$$\tan(4°) = \frac{\text{opposite}}{\text{adjacent}} = \frac{1000}{x}$$

$x \cdot \tan(4^\circ) = 1000$

$x = \dfrac{1000}{\tan(4^\circ)} \approx 14{,}300.66626$ ft.

The ship is about 14,300 feet from shore.

9. $\cos(19^\circ) > \cos(20^\circ)$

Recall that the cosine of an angle is equal to the x-coordinate of the intersection of the angle's terminal side with the unit circle. Studying an exaggerated diagram of a portion of the unit circle, shows that the cosine of 19° (the x-coordinate) is greater than the cosine of 20°.

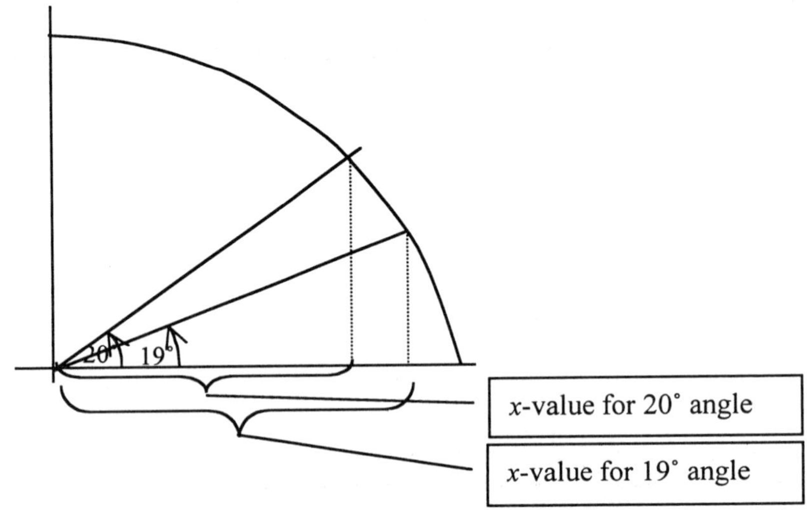

10. There are several different ways to find AB if additional information were available. The chart below summarizes the possibilities:

additional information	method for finding AB
AC	Pythagorean Theorem
$\angle B$	cosine ratio
$\angle C$	sine ratio
AB	then we're done.

Lesson 6 Trigonometry in Construction

Homework

1.

grade = 7%
slope = 0.07

3000 ft

grade = 9%
slope = 0.09

Town A M Town B

The tunnel will go along the straight line from Town A through the point M and then to Town B. The distance from Town A to point M:

$\text{slope} = \dfrac{\text{rise}}{\text{run}}$

$0.07 = \dfrac{3000 \text{ ft}}{\text{run}}$

run = 3000/0.07 ≈ 42,857.14 feet

The distance from M to Town B:

$$\text{slope} = \frac{\text{rise}}{\text{run}}$$

$$0.09 = \frac{3000 \text{ ft}}{\text{run}}$$

$\text{run} = 3000/0.09 \approx 33{,}333.33$ feet

Adding the two distances yields the tunnel length

$42{,}857.14 + 33{,}333.33 = 76{,}190.47$ feet

converting the distance to miles yields

$76{,}190.47 \cdot \dfrac{1 \text{ mile}}{5280 \text{ feet}} = 14.43$ miles for the length of the tunnel between Town A and Town B.

2.

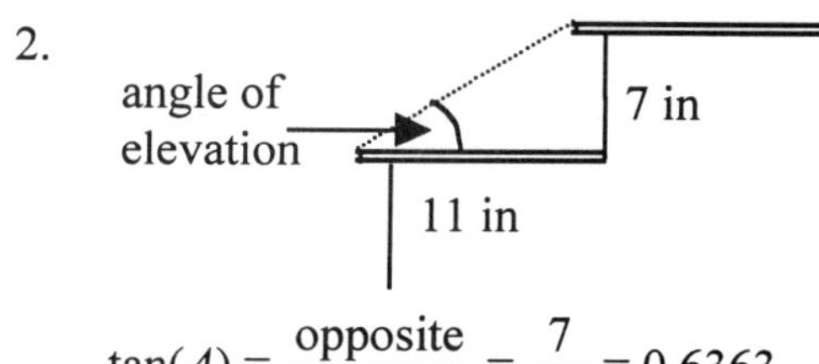

$$\tan(A) = \frac{\text{opposite}}{\text{adjacent}} = \frac{7}{11} = 0.6363$$

$\tan^{-1}(0.6363) \approx 32.47^\circ$. The angle of elevation of a typical staircase is 32.47°.

Lesson 7 A Laboratory of Indirect Measure

Lesson 8 Applications in Astronomy and Earth Science

Homework

2. b. circumference = $\pi \cdot$diameter

$39{,}780 = \pi \cdot \text{diameter}$

$\text{diameter} \approx 12{,}662.3673$ km.

4. $$\frac{\text{actual distance}}{2\pi \cdot \text{distance}} = \frac{\text{angular distance}}{360^\circ}$$

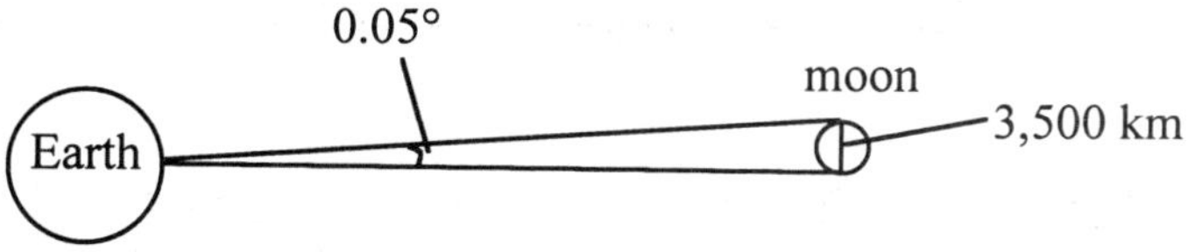

$$\frac{3500}{2\pi \cdot \text{distance}} = \frac{0.5}{360^\circ}$$

$\text{distance} \approx 401{,}070$ km

The actual distance from the Earth to the Moon varies from 356,000 km to 407,000 km. Our estimate was pretty good.

Sampling

Lesson 1 Census versus Survey

Homework

1. a. people who eat dinner out 4 or more times per week; b. price per person per dinner.
   or a. vegetables; b. percent of vitamin a, b, c, etc. per serving.
   or a. people who are vegetarians; b. number of years they've been vegetarian.

3. a. people employed full time; b. number of vacation days per year.
   or a. people who took a vacation last year; b. percentage who went to a colder climate.

5. a. people employed full time at a particular company; b. number who choose HMO versus non-HMO coverage. or a. people who live in a particular state; b. the number with comprehensive health insurance.

Lesson 2 Sample Error

Homework

1. c. Ask question so as to mimic voting process; Holding wording of question constant; Test several wordings of new questions.

5. People surveyed about an election don't actually vote in the election; or the election has unusually low voter turnout. In addition, people may not want to admit that they do not understand a question.

Lesson 3 Summarizing Our Product's Characteristics

Homework

Answers will vary. The following are examples.

1. Shelf life of canned food; the percent of calories from fat.

3. In health clubs, the availability of workout machines; the percentage of members that are female.

5. The number of clothes items returned due to faulty merchandise; the number of socks without holes after 3 months of use.

9. Movies and the difference between the highest and lowest acting salary; the average amount of money taken in at an amusement park per day.

Lesson 4 The Normal Approximation to a Binomial Distribution

Homework

1. a. $\mu = n{\cdot}p = 10{\cdot}(0.76) = 7.6$. The mean number of free throws of 10 free throw attempts is 7.6.
   c. $\mu - \sigma = 7.6 - 1.35 = 6.25$. This rounds up to 7.
   $\mu + \sigma = 7.6 + 1.35 = 8.95$. This rounds down to 8.

3. a. One approach is to find the average score of all the golfers. Add all the scores and divide by the total number of golfers.

$$\frac{63+64+65+4\cdot66+2\cdot67+9\cdot68+15\cdot69+17\cdot70+17\cdot71+16\cdot72+25\cdot73+10\cdot74+12\cdot75+4\cdot76+1\cdot77+2\cdot78}{137}$$

$$= \frac{9788}{137} \approx 71.5$$

Another approach is to make a histogram of the golf scores and draw the bell-curve over that histogram. Since the top of the bell curve occurs at μ, we can read the value of μ from the graph, or at about a score of 71 or 72.
   c. Since 69 and 74 both differ from μ (71.5) by 2.5, a good value for σ would be 2.5.
   e. $\mu - 3\sigma = 71.5 - 3(2.5) = 71.5 - 7.5 = 64$, rounded up to 64
   $\mu + 3\sigma = 71.5 + 3(2.5) = 71.5 + 7.5 = 79$, rounded down to 79
4. c. $\mu - \sigma = 9 - 0.95 = 8.05$, rounded up to 9
   $\mu + \sigma = 9 + 0.95 = 9.95$ rounded down to 9
   d. $\mu - \sigma$, $\mu + \sigma$ specifies a range that includes a score of 9. Table 6 includes 10 frames with scores of 9. 10 out of 30 frames is 33.3%
   g. The bowlers' "number of pins knocked down per frame" doesn't appear to be approximately normal. The number of frames within $\mu \pm \sigma$ represents only 33.3% of the frames. In approximately normal distributions about 68% of the frames would be within $\mu \pm \sigma$.

Lesson 5 Characteristics of Samples

Homework

1. From problem 3 of Lesson 4 homework, $\mu = 71.5$ and $\sigma = 2.5$
   a. The variance of a population: $\sigma^2 = (2.5)^2 = 6.25$
   b. The variance of sample means: $S^2 = \frac{\sigma^2}{n} = \frac{6.25}{4} = 1.5625$
   The standard deviation of sample means: $S = \sqrt{S^2} = \sqrt{1.5625} = 1.25$

Lesson 6 Control Charts

Homework

1. The different colors weren't well mixed. An employee may forget to add a particular color or may add a color twice. A machine malfunction may prevent addition of a color or cause too many of a particular color to be added. All the M&M's may melt together into one large mud-colored M&M.

2. c. Percent of female members of health clubs; opening of a women's-only club nearby; a promotion to attract female members; canceling of daycare services.
   e. number of socks without holes after 3 months; change in fibers used to manufacture socks.
   i. difference between highest and lowest acting salaries; actors labor union legal action.

Lesson 7 SPC in a Semiconductor Factory

Homework

1. a. If the sole is less than $11\frac{3}{8} - \frac{1}{4} = 11\frac{1}{8}$ inches or greater than $11\frac{3}{8} + \frac{1}{4} = 11\frac{5}{8}$ inches, the sole would be considered defective.
   b. The customer has stated that the desired sole length is $11\frac{3}{8}$ inches, so $\mu = 11\frac{3}{8}$.

4. a. $2\sigma + 2\sigma = (\mu - \text{LL}) + (\text{UL} - \mu)$

$$4\sigma = (11\frac{3}{8} - 11\frac{1}{8}) + (11\frac{3}{8} - 11\frac{3}{8})$$

$$4\sigma = \frac{1}{4} + \frac{1}{4}$$

$$4\sigma = \frac{1}{2}$$

$$\sigma = \frac{1}{8}$$

5. b. cost = (cost per shoe) (number made)
   cost = (10) (10,000) = \$100,000
   Before computing the number sold, you must recall that a manufacturing process has ±3 sigma defect levels, fitting the 99% rule, so approximately 1% of the product will be defective. The defective shoes aren't sold.
   number sold = number made – defects
   number sold = 10,000 – (1% of 10,000)
   number sold = 10,000 – 100
   number sold = 9900
   revenue = (price per shoe) (number sold)
   revenue = (12.50) (9900)
   revenue = \$123,750
   profit = revenue – cost
   profit = 123,750– 100,000
   profit = \$23,750.
   The profit grows as the business has fewer defective products.

Lesson 8 How Defects Affect Profitability

Homework

2. a. Find the new cost per M&M
   cost per M&M + 0.75(cost per M&M) = price per M&M
   $1.75 \cdot$ cost per M&M = 0.0045

cost per M&M = 0.00257

Find the first ordered pair, the maximum profit (no defects)

cost = (unit cost)(quantity made)

$= 0.00257 \cdot 100 = 0.257$

revenue = (unit price)(quantity sold)

$= 0.0045 \cdot 100 = 0.45$

profit = revenue – cost

$= 0.45 - 0.257$

$= 0.193,$

The ordered pair is (0, 0.193)

Find the second ordered pair, (20 defects)

cost = (unit cost)(quantity made)

$= 0.00257 \cdot 100 = 0.257$

revenue = (unit price)(quantity sold)

$= 0.0045 \cdot 80 = 0.36$

profit = revenue – cost

$= 0.36 - 0.257$

$= 0.103,$

The ordered pair is (0, 0.103)

Find the linear model.

$$m = \frac{P_2 - P_1}{d_2 - d_1} = \frac{0.103 - 0.193}{20 - 0} = -0.0045$$

$P - P_1 = m(d - d_1)$

$P - 0.193 = -0.0045(d - 0)$

$P - 0.193 = -0.0045d$

$P = -0.0045d + 0.193$ Profit is expressed as a function of the number of defects.

b. Find the new cost per M&M

cost per M&M + 0.75(cost per M&M) = price per M&M

$1.25 \cdot$ cost per M&M = 0.0045

cost per M&M = 0.0036

Find the first ordered pair, the maximum profit (no defects)

cost = (unit cost)(quantity made)

$= 0.0036 \cdot 100 = 0.36$

revenue = (unit price)(quantity sold)

$= 0.0045 \cdot 100 = 0.45$

profit = revenue – cost

$= 0.45 - 0.36$

$= 0.09,$

The ordered pair is (0, 0.09)

Find the second ordered pair, (20 defects)

cost = (unit cost)(quantity made)

$= 0.0036 \cdot 100 = 0.36$

revenue = (unit price)(quantity sold)

$= 0.0045 \cdot 80 = 0.36$

profit = revenue – cost

$= 0.36 - 0.36$

$= 0.00,$

The ordered pair is (0, 0)

Find the linear model.

$$m = \frac{P_2 - P_1}{d_2 - d_1} = \frac{0.0 - 0.09}{20 - 0} = -0.0045$$
$P - P_1 = m(d - d_1)$
$P - 0.09 = -0.0045(d - 0)$
$P - 0.09 = -0.0045d$
$P = -0.0045d + 0.09$ Profit is expressed as a function of the number of defects.

4. d. If you think that you will drink at least 5 cups of coffee, it's wiser to buy the mug. Besides, you'll get a free cup after 5 cups for no additional cost. If you think you will drink fewer than 5 cups and you don't want the mug, then you should buy the coffee by the cup.

Lesson 9 Deming's All or Nothing Acceptance Sampling Scheme

Homework

1. For a. – f. answers will vary. You should consider the nature of the product and the severity of the consequences due to defective parts. Some of the products are almost totally man-made (e.g. electronic parts to a cell phone company), while others are not man-made (e.g. grain). Some of the products could result in very dire consequences (e.g. many deaths) if they were defective, while others may result in milder consequences (e.g. stomachache).
   a. Defects in grain could include mold, infestation or excessively early harvesting. Defective grain may only cause minor problems and can sometimes be used for other purposes (e.g. seed). Accordingly, the p value may be relatively high.
   b. Defects in flour could include infestation, mold or other inappropriate milling. Defective flour could contaminate entire batches of bread. Contaminated bread may need to be discarded causing economic loss. If the contamination is not detected, it could result in health problems when consumed. Accordingly, the p value for flour may be lower than the p value for grain, but it is probably still higher than the p value for airplane turbine engines.
   c. There are multiple causes for defective turbine engines. Such engines consist of many parts, any one of which could be defective. In addition to the number of parts, the assembly of the engine could be defective. While there are many sources of defects in turbine engines, the consequences of such defects are quite severe. Turbine engines are expensive to dismantle, test, repair and reassemble. Thus, a defective engine can cause economic loss. If a defective engine is not detected, it could result in a plane crash and major loss of life. Accordingly the p value for airplane turbine engines should be very small.
   d. Defects to oxygen tanks might include improper pressure, defective valves and contaminated oxygen. The number of parts and complexity of the assembly are less than in a turbine engine. A defective oxygen tank could lead to loss of life. Accordingly, the p value for oxygen tanks should be lower than for grain or flour, but could probably be higher than the p value for airplane turbine engines.
   e. Electronic parts can have a number of defects. The parts of the assembly could themselves be defective or defects could have occurred as the product was assembled. However, many electronic parts are no longer "assembled" as much as they are "grown." This manufacturing significantly reduces the number of defects in electronic parts. This would suggest a rather low p value. On the other hand, the consequences of a defective part in a cell phone are relatively minor. Dismantling, repairing and reassembling a cell phone is a relatively quick and easy task (as compared to the turbine engine). Also, a malfunctioning cell phone is most likely to be a minor irritant, not a cause of death.
   f. Motors have a number of parts, each of which might be defective. In addition, assembly of the motor could be defective. Thus, there are a number of sources of defects. The failure of a motor could lead to economic loss due to the cost of disassembly, repair, and reassembly.

The failure of an appliance, due to a faulty motor, would be an inconvenience and may lead to the loss of a customer, but loss of life or a significant financial loss is unlikely. Accordingly, the p value for a motor is probably relatively high as compared to p values for oxygen tanks or phone parts.

4. From the information provided in the question, $p = \dfrac{1}{1000}$, $k_1 = 0.25$, and $k_2 = 500$

   $\dfrac{k_1}{k_2} = \dfrac{0.25}{500}$ or $\dfrac{1}{2000}$ Since $p > \dfrac{k_1}{k_2}$, this is a Case 2 situation and the business should inspect each transaction.

5. c. k is the average cost to test one or more parts to find a good one in the surplus supply S. A formula for k is k = (cost for 1st test) + (cost for 2nd test) $\cdot\, p$ + (cost for 3rd test) $\cdot\, p^2 + \ldots$

   $= k_1 + k_1 \cdot p + k_1 \cdot p^2 + k_1 \cdot p^3 + k_1 \cdot p^4 + \ldots$

   If k_1 is factored out, $k = k_1(p + p^2 + p^3 + p^4 + \ldots)$

   Parts a and b of this problem suggest that $\dfrac{1}{1-p} = p + p^2 + p^3 + p^4 + \ldots$

   So, the formula becomes: $k = k_1 \cdot \dfrac{1}{1-p}$

   $1 - p$ is the average fraction of good parts, also known as q.

   $k = k_1 \cdot \dfrac{1}{q} = \dfrac{k_1}{q}$

   Thus, we have shown that the average cost to test one or more parts to find a good one in the surplus supply is $\dfrac{k_1}{q}$.

Sets and Logic

Lesson 1 Set for Life

Homework

3. The following sets are subsets of {2, 6, 8}: {2, 6, 8}, {2, 6}, {2, 8}, {6, 8}, {2}, {6}, {8}, ∅.

8. The elements of $A \cap B$
   {1, 2, 3, 4, 5, } ∩ {3, 4, 5, 6, 7, 8}
   {3, 4, 5}

9. The elements of $A \cup B$
   {1, 2, 3, 4, 5, } ∪ {3, 4, 5, 6, 7, 8}
   {1, 2, 3, 4, 5, 6, 7, 8} or U

13. The elements of $A' \cap B$
   $A' = \{6, 7, 8\}$ and $B = \{3, 4, 5, 6, 7, 8\}$
   $A' \cap B = \{6,7,8\} = A'$

16. The elements of $(A \cap B)'$
   $A = \{1, 2, 3, 4, 5\}$ and $B = \{3, 4, 5, 6, 7, 8\}$ and $U = \{1, 2, 3, 4, 5, 6, 7, 8\}$
   $(A \cap B)'$
   $(\{1, 2, 3, 4, 5\} \cap \{3, 4, 5, 6, 7, 8\})'$
   $(\{3, 4, 5\})'$
   {1, 2, 6, 7, 8}

17. The elements of $(A \cup B) \cap B'$
   $A = \{1, 2, 3, 4, 5\}$ and $B = \{3, 4, 5, 6, 7, 8\}$ and $U = \{1, 2, 3, 4, 5, 6, 7, 8\}$
   $(A \cup B) \cap B'$
   $(\{1, 2, 3, 4, 5\} \cup \{3, 4, 5, 6, 7, 8\}) \cap (\{3, 4, 5, 6, 7, 8\})'$
   $(\{1, 2, 3, 4, 5, 6, 7, 8\} \cap (\{1, 2\})'$
   {1, 2} or B

Lesson 2 Picture This

Homework

4. $A \cap B \cap C = \{30\}$

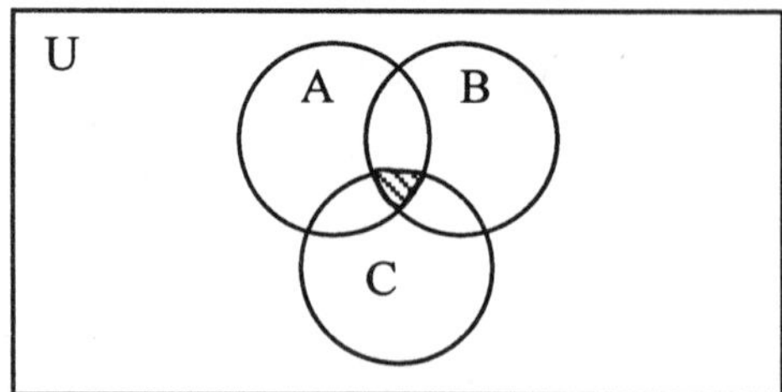

7. $(A \cap C)'$
$= \{5, 10, 20, 25, 35, 40, 45, 50, 55, 60\}$

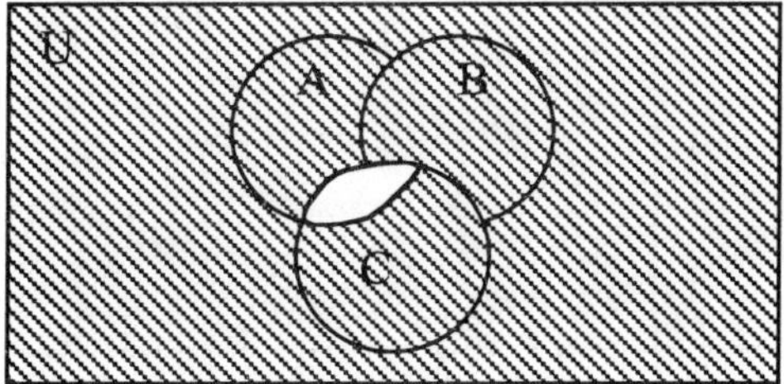

17. $A \cap (B \cup C)$

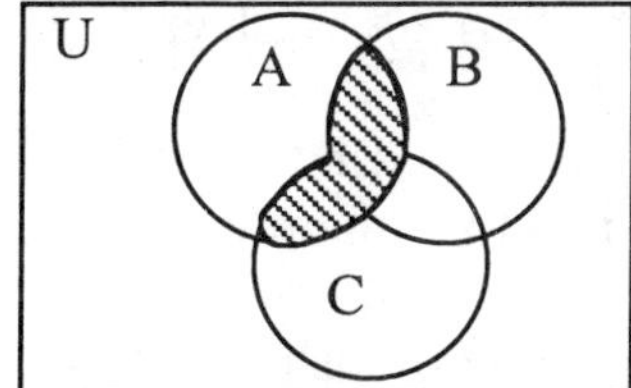

20. $(A \cap B) \cup C$

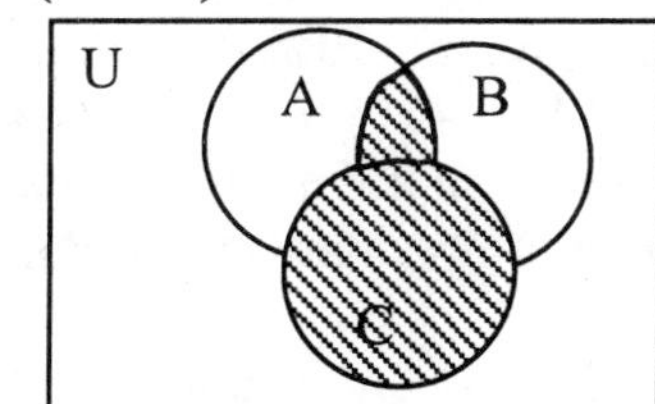

Lesson 3

Homework

2.

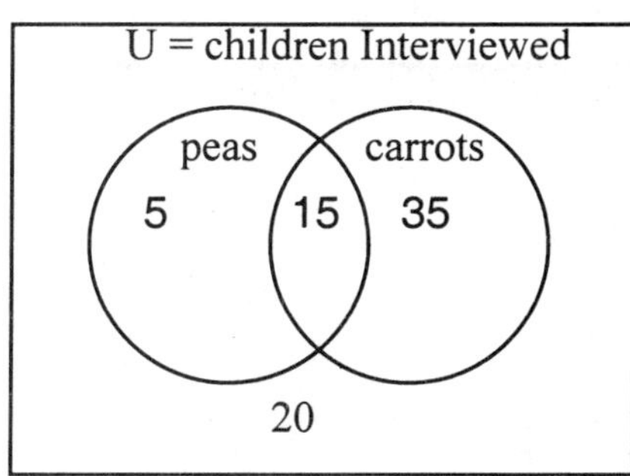

a. How many children like peas, but not carrots?
children interviewed – (# like only carrots + # like both peas/carrots + like neither)
75 – (20 + 35 + 16) = 75 – 70 or 5 who like only carrots.

b. How many children like carrots?
like only carrots + # like peas/carrots = # like carrots
35 + 15 = 50 children who like carrots.

c. How many children like peas?
like only peas + # like peas/carrots = # like peas
5 + 15 = 20 children who like peas.

4. Draw a Venn diagram and find the following:

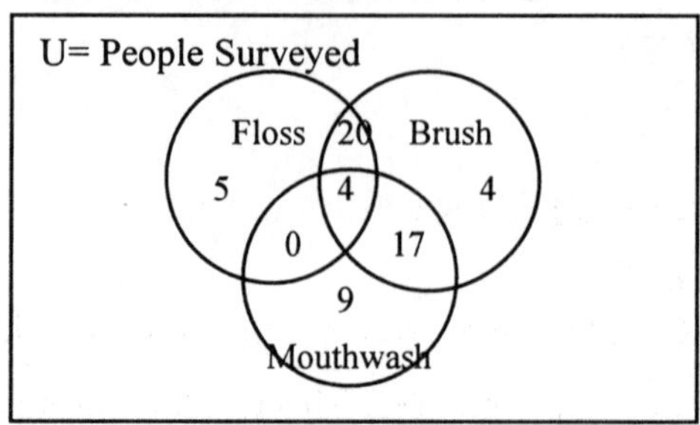

a. The number of people who brush their teeth, but do not floss their teeth or use mouthwash.
brush – (# brush/floss + # brush/floss/mouthwash + # brush/mouthwash) = # brush only
45 – (20 + 4 + 17) = 45 – 41 or 4 people brush only.

b. The number of people who use mouthwash and floss their teeth, but do not brush their teeth.
mouthwash – (# mouthwash/brush + # mouthwash/brush/floss – # mouthwash only) = # mouthwash/floss
30 – (17 + 4 + 9) = 30 – 30 or 0 people who floss and mouthwash but do not brush their teeth.

c. The number of people who use mouthwash, but do not floss their teeth or brush their teeth 9
mouthwash – (# mouthwash/floss + # mouthwash/rush + # mouthwash/brush/floss) = # mouthwash only.
30 – (0 + 17 + 4) = 30 – 21 or 9 people who use mouthwash, but neither floss nor brush their teeth.

Lesson 4 Count Your Chips

Homework

2. A pair of dice is rolled and the top faces are recorded.

{ 11, 12, 13, 14, 15, 16,
21, 22, 23, 24, 25, 26,
31, 32, 33, 34, 35, 36,
41, 42, 43, 44, 45, 46,
51, 52, 53, 54, 55, 56,
61, 62, 63, 64, 65, 66,}

4. The sample space is the single number 1-4, that is selected each time.

{ 11, 12, 13, 14,
21, 22, 23, 24,
31, 32, 33, 34,
41, 42, 43, 44 }

9. The elements where there is at least one Tail: {HT, TH, TT}

13. The elements where at least one defective calculator is found: All except GGGG.

16. The elements where both officers selected are female: {Mary/Jane, Jane/Mary, Mary/Sally, Sally/Mary, Jane/Sally, Sally/Jane}.

Lesson 5 Learning to Reason

Homework

1. It is sunny outside.
   Yes, it is a proposition because we can decide whether it is sunny or not.

3. Chevrolet makes a better truck than Ford.
   No. "Better" is not something that can be proven true or false. It is an opinion.

6. There are more integers than rational numbers. Yes. This is a false proposition.

8. The area of the square is 64 and the perimeter is 32. $p \wedge q$

12. This homework is either hard or it is easy.
    $p \vee q$ The homework is hard OR the homework is easy.

16. $\sim(\sim p) \vee q$

p	q	$\sim p$	$\sim(\sim p)$	$\sim(\sim p) \vee q$
T	T	F	T	T
T	F	F	T	T
F	T	T	F	T
F	F	T	F	F

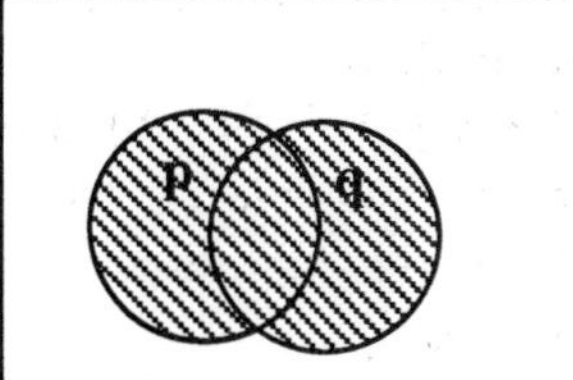

16. Alternate solution. Since $\sim(\sim p) = p$, $\sim(\sim p) \vee q = p \vee q$

p	q	$p \vee q$
T	T	T
T	F	T
F	T	T
F	F	F

18. $(p \wedge q) \wedge r$

p	q	r	$p \wedge q$	$(p \wedge q) \wedge r$
T	T	T	T	T
T	T	F	T	F
T	F	T	F	F
T	F	F	F	F
F	T	T	F	F
F	T	F	F	F
F	F	T	F	F
F	F	F	F	F

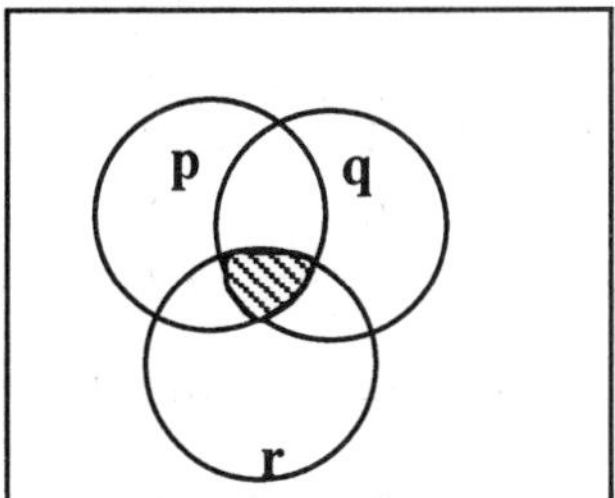

Lesson 6 Propositional Calculus

Homework

2. $(p \vee q) \vee \sim p$

	q	$p \vee q$	$\sim p$	$(p \vee q) \vee \sim p$
T	T	T	F	T
T	F	T	F	T
F	T	T	T	T
F	F	F	T	T

$p \vee q$

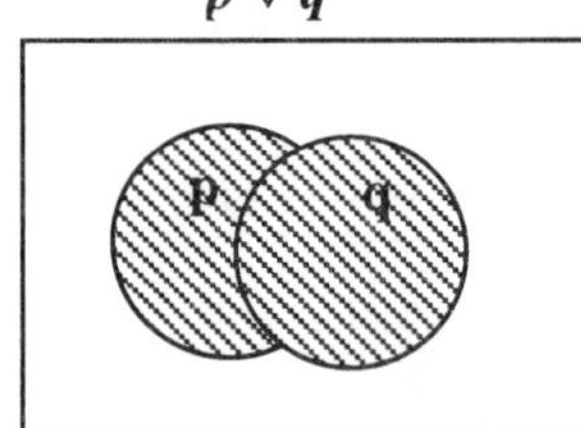

$\sim p$

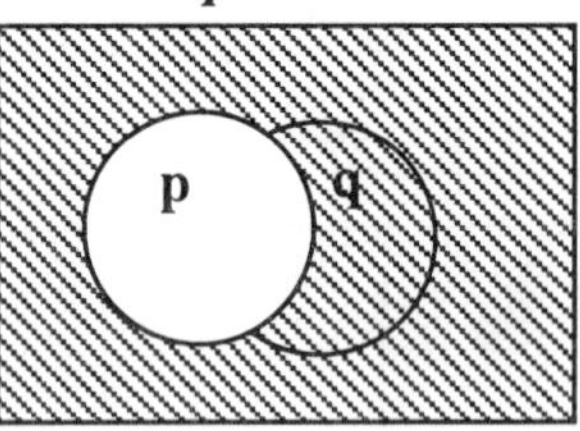

$(p \vee q) \vee \sim p$

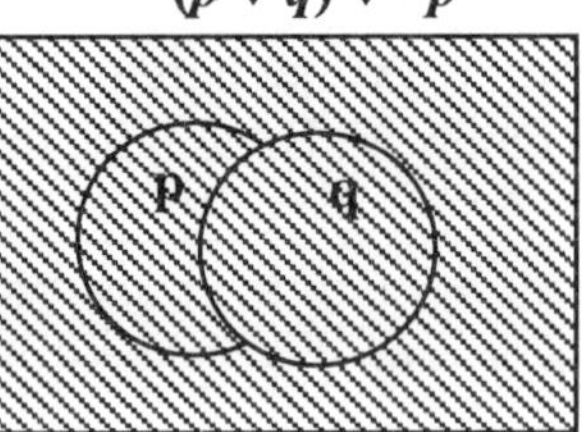

7. $(p \wedge q) \wedge r \equiv p \wedge (q \wedge r)$

p	q	r	$p \wedge q$	$q \wedge r$	$(p \wedge q) \wedge r$	$p \wedge (q \wedge r)$
T	T	T	T	T	T	T
T	T	F	T	F	F	F
T	F	T	F	F	F	F
T	F	F	F	F	F	F
F	T	T	F	T	F	F
F	T	F	F	F	F	F
F	F	T	F	F	F	F
F	F	F	F	F	F	F

8. $(p \vee q) \vee r \equiv p \vee (q \vee r)$

p	q	r	$p \vee q$	$q \vee r$	$(p \vee q) \vee r$	$p \vee (q \vee r)$
T	T	T	T	T	T	T
T	T	F	T	T	T	T
T	F	T	T	T	T	T
T	F	F	T	F	T	T
F	T	T	T	T	T	T
F	T	F	T	T	T	T
F	F	T	F	T	T	T
F	F	F	F	F	F	F

Lesson 7 Logical Applications

Homework

1. a. I am not going to Burger King or McDonald's or Wendy's for lunch is equivalent to I am not going to Burger King and I am not going to McDonald's and not Wendy's.

b. Construct a truth table to prove this extension.

p	q	r	$\sim p$	$\sim q$	$\sim r$	$q \vee r$	$p \vee (q \vee r)$	$\sim[p \vee (q \vee r)]$	$\sim q \wedge \sim r$	$\sim p \wedge (\sim q \wedge \sim r)]$
T	T	T	F	F	F	T	T	F	F	F
T	T	F	F	F	T	T	T	F	F	F
T	F	T	F	T	F	T	T	F	F	F
T	F	F	F	T	T	F	T	F	T	F
F	T	T	T	F	F	T	T	F	F	F
F	T	F	T	F	T	T	T	F	F	F
F	F	T	T	T	F	T	T	F	F	F
F	F	F	T	T	T	F	F	T	T	T

6. $p \wedge (q \vee r) \Rightarrow (p \wedge q) \vee (q \wedge r)$

p	q	r	$q \vee r$	$p \wedge (q \vee r)$	$p \wedge q$	$q \wedge r$	$(p \wedge q) \vee (q \wedge r)$	$p \wedge (q \vee r) \Rightarrow (p \wedge q) \vee (q \wedge \text{r})$
T	T	T	T	T	T	T	T	T
T	T	F	T	T	T	F	T	T
T	F	T	T	T	F	F	F	F
T	F	F	F	F	F	F	F	T
F	T	T	T	F	F	T	T	F
F	T	F	T	F	F	F	F	T
F	F	T	T	F	F	F	F	T
F	F	F	F	F	F	F	F	T

Lesson 8 Seek and You Shall Find

Homework

2. not (Number of Years of Education = 12): John Jones, Jim Smith, Sally White, Pat Brown, Bill Jones

9. not ((Home State = Arizona) and (Number of Years of Education = 20))
   John Jones, Jim Smith, Pat Brown, Karen Black, Bill Jones, Bob Clark

Lesson 9 Computer Logic

Homework

2. The symbolic equivalent is: $\sim((p \vee \sim p) \wedge (\sim q))$

p	q	$\sim p$	$p \vee \sim p$	$\sim q$	$(p \vee \sim p) \wedge \sim q$	$\sim((p \vee \sim p) \wedge \sim q)$
1	1	0	1	0	0	1
1	0	1	1	1	1	0
0	1	0	1	0	0	1
0	0	1	1	1	1	0

Notice that this logic gate network gives the same result as the "more complicated logic gate network" example given in the lesson.

4. The symbolic equivalent is: $p \vee (\sim q \wedge \sim r)$

p	q	r	$\sim q$	$\sim r$	$\sim q \wedge \sim r$	$p \vee (\sim q \wedge \sim r)$
1	1	1	0	0	0	1
1	1	0	0	1	0	1
1	0	1	1	0	0	1
1	0	0	1	1	1	1
0	1	1	0	0	0	0
0	1	0	0	1	0	0
0	0	1	1	0	0	0
0	0	0	1	1	1	1

Lesson 10 More or Less

Homework

1. $x \leq -4$ Include all numbers that are less than or equal to –4.

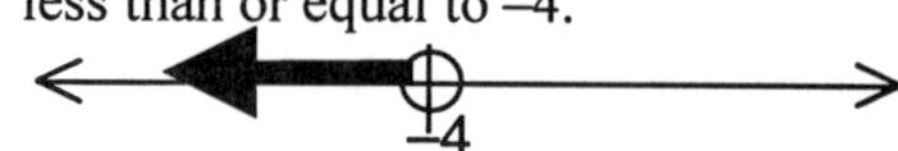

2. $w \geq 19$ Include all numbers that are greater than or equal to 19.

3. $t \geq -2$ Include all numbers that are greater than or equal to –2.

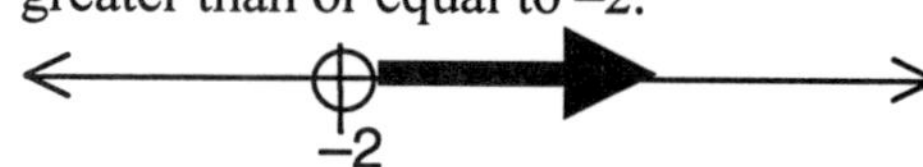

4. $2d + 3 > 9$
   $2d > 6$ subtract 3 from both sides
   $d > 3$ divide both sides by 2

5. $-2t - 6 < 12$
   $-2t < 18$ add 6 to both sides
   $t > -9$ divide both sides by –2, reversing the direction of the inequality.

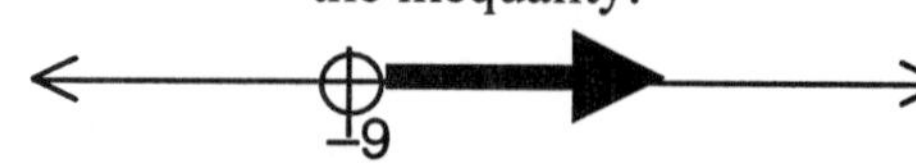

6. $-g + 4 > -5$
   $-g > -9$ subtract 4 from both sides
   $g < 9$ multiply both sides by –1 reverse the direction of the inequality

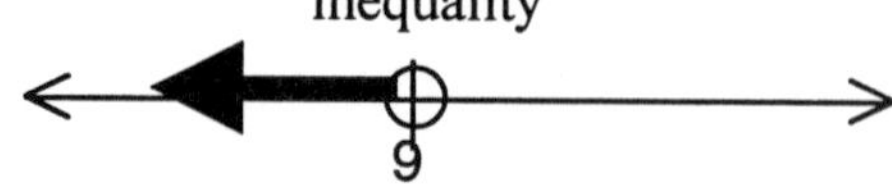

7. a. \$30 per hour ($30t$) plus a \$20 flat fee (20) must be less than or equal to \$100.
   In equation form, $30t + 20 \leq 100$

   b. $30t + 20 \leq 100$

   $30t \leq 80$ subtract 20 from both sides

   $t \leq \frac{80}{30}$ divide both sides by 30

   $t \leq \frac{8}{3}$ simplify the fraction

   $t \leq 2\frac{2}{3}$ Write as a mixed number, interpreting the result as 2 and 2/3 hours.

   We can write $0 \leq t \leq 2\frac{2}{3}$, since negative hours worked would have no meaning.

   c. The customer can afford the plumber to work between 0 hours and 2 hours, 40 minutes.

8. Let M represent the width of a wire. Its target width is 2 microns.

   $M > 2$ microns – 0.005 microns

   $M < 2$ microns + 0.005 microns

   This can be written as: $2 - 0.005 < M < 2 + 0.005$

   $1.995 < M < 2.005$ simplify the inequality

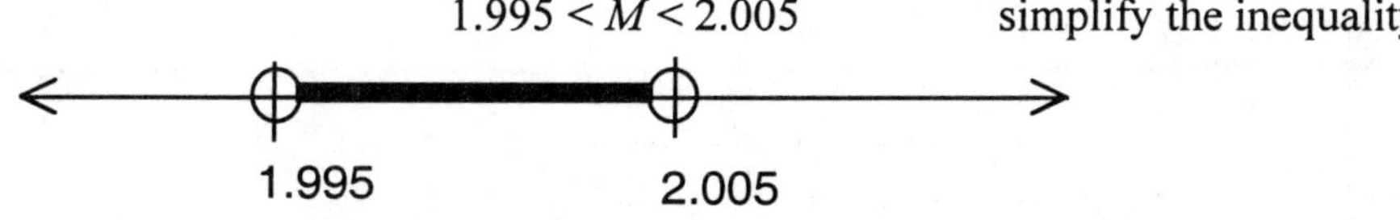

This is a conjunction because the word AND is used and also because the graph is shaded between the two endpoints.

Systems

Lesson 1 Lake Victoria

Homework

1. The predators cannot consume the entire prey population and cause the prey to become extinct, because as soon as the demand for prey exceeds the supply, some predators will begin to die of starvation, thus reducing the numbers of prey consumed, so that all will not be eaten up.

   Another refinement of the argument: since a predator consumes many times more than its body weight in prey, when one predator dies due to insufficient food, the number of prey should increase significantly so that they could never become extinct.

6. a. $3x + 19 = 35$
   $-19 \quad -19$
   $3x = 16$
   divide by 3
   $x = 16/3 \approx 5.333$

   b. $13x - 1 = 9x + 14$
   $-9x \quad -9x$
   $4x - 1 = 14$
   $+1 \quad +1$
   $4x = 15$
   divide by 4
   $x = 15/4 = 3.75$

   c. $3.1y - 9.6 = 1.2$
   $+9.6 \quad +9.6$
   $3.1y = 10.8$
   divide by 3.1
   $y \approx 3.48$

   d. $5.7a - 12 = 2.1a - 1$
   $-2.1a \quad -2.1a$
   $3.6a - 12 = -1$
   $+12 \quad +12$
   $3.6a = 11$
   divide by 3.6
   $a \approx 3.056$

   e. $\frac{5x}{8} = 3$
   multiply by 8/5
   $x = 24/5 = 4.8$

   f. $\frac{2x}{3} = \frac{3x}{4} + 1$
   multiply by 12
   $8x = 9x + 12$
   $-9x \quad -9x$
   $-x = 12$
   multiply by -1
   $x = -12$

7. a. $5 + 3 \cdot 8 - 10$
   $= 5 + 24 - 10$
   $= 29 - 10$
   $= 19$

   b. $2 - 6(2 - 5)$
   $= 2 - 6 \cdot (-3)$
   $= 2 + 18$
   $= 20$

   c. $2(5 \cdot 2 - 18) - 3$
   $= 2(10 - 18) - 3$
   $= 2(-8) - 3$
   $= -16 - 3$
   $= -19$

   d. $3 - 2(5 \cdot 2 - 18)$
   $= 3 - 2(10 - 18)$
   $= 3 - 2(-8)$
   $= 3 + 16$
   $= 19$

Lesson 2 Cost and Revenue

Nitty Gritty:

1. $x + 4y = 3$
   $-x \quad -x$
   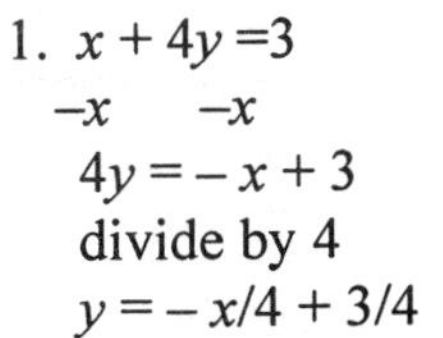
   $4y = -x + 3$
   divide by 4
   $y = -x/4 + 3/4$

   or
   $m = -\text{A/B} = -1/4$
   $b = \text{C/B} = 3/4$
   $y = (-1/4)x + 3/4$

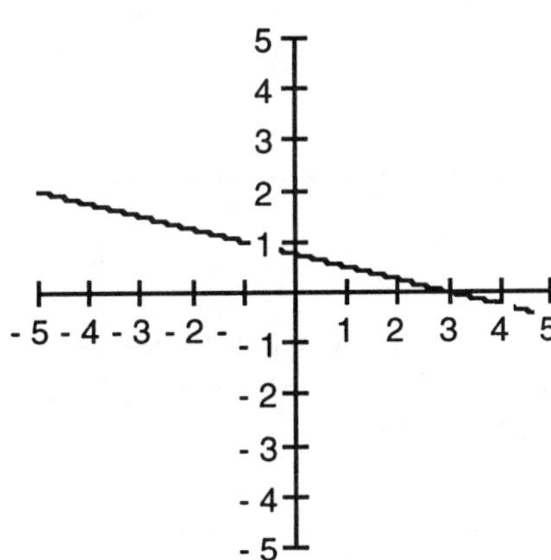

3. $5x - 2y = 4$
   $-5x \quad -5x$
   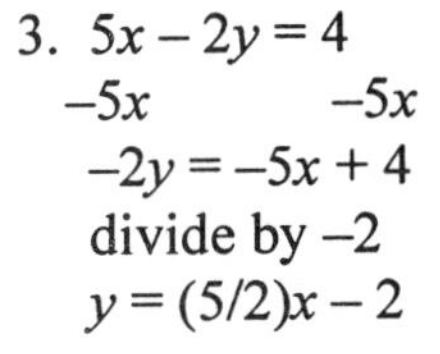
   $-2y = -5x + 4$
   divide by -2
   $y = (5/2)x - 2$

   or
   $m = -\text{A/B} = (-5)/(-2) = 5/2$
   $b = \text{C/B} = 4/(-2) = -2$
   $y = \frac{5}{3}x - 2$

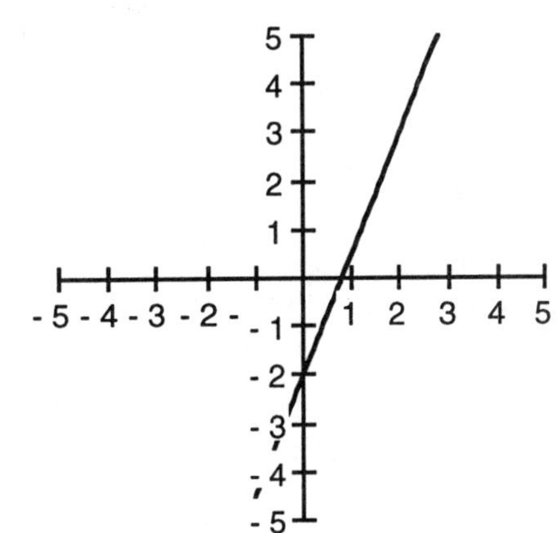

Homework

2. Graph **a** in problem 1 does not represent a realistic situation because it indicates that when the catch size is zero, the income is not zero. This says that we made money from catching nothing!

   Graph **b** in problem 1 does not represent a realistic situation because again, for zero catch size, there was still an income. In addition, once the catch size begins to increase, the income level decreases, and the expenses also decrease. This is the opposite of the real world scenario. Finally, once a certain catch size is reached, from then on there are no more expenses, yet the income level continues to decrease.

   Graph **c** in problem 1 does not represent a realistic situation because as in the two previous cases, there is an income from the catch despite a catch size of zero. Expenses should not decrease, as the catch size gets larger. (Cost per fish might get smaller, though.)

   Graph **d** in problem 1 does not represent a realistic situation because there is an income from a zero catch size. Other than that, it is feasible, although not expected, that expenses might increase faster than income with increasing catch size. This simply indicates that catching this kind of fish is not a money making proposition, because the income is too small.

5. a. First solve for y in both equations to get $y = (3/5)x - 11/5$ and $y = x/3 - 5/3$.

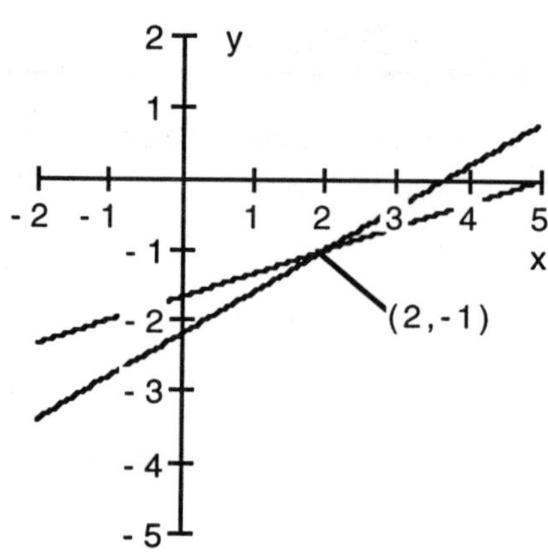

b. First solve for y in both equations to get $y = -x/2 + 5/2$ and $y = x/3 + 5/3$.

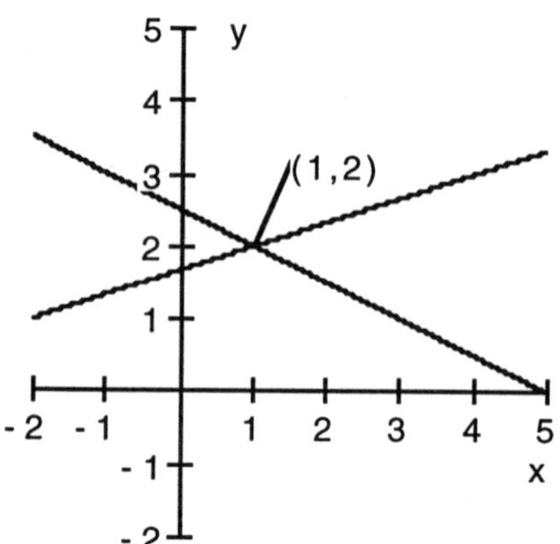

8. Let p = Price per thousand fish
1500 = Cost
$100p$ = Revenue
At break-even, Cost = Revenue.
$1500 = 100p$
divide by 100
$15 = p$
The fisherman must get $15 per thousand fish to break-even.

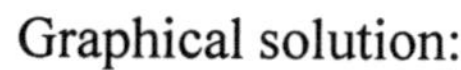
Graphical solution:

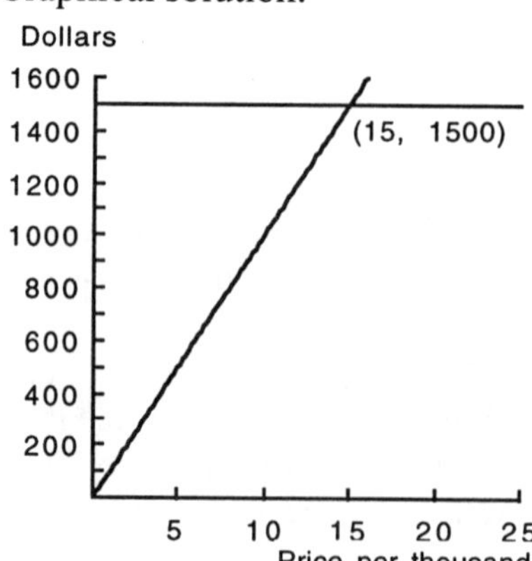

11. Let x = Number of fish caught in thousands
y = Cost
m = Price per thousand fish
$m \cdot x$ = Revenue
To break even, Cost = Revenue.
$y = m \cdot x$

"m" is the slope of the line in some of the standard forms for an equation of a line.

Lesson 3 Linear Models

Nitty Gritty: Equation of a Line Through Two Points

1. $5 = 2a + b$ or $b = 5 - 2a$
$8 = a + b$
Substituting into the second equation:
$8 = a + (5 - 2a)$
$8 = 5 - a$
$+a \quad +a$
$8 + a = 5$
$-8 \quad -8$
$a = -3$
Solving for b:
$5 - 2(-3) = b$
$b = 5 + 6 = 11$
The equation is $y = -3x + 11$

3. $7 = 6a + b$ or $b = 7 - 6a$
$-5 = -2a + b$
Substituting into the second equation:
$-5 = -2a + (7 - 6a)$
$-5 = -8a + 7$
$-7 \quad -7$
$-12 = -8a$
divide by -8
$a = (-12)/(-8) = 3/2$
Solving for b:
$10 + 2\cdot(-5/2) = b$
$b = 7 - 6(3/2) = 7 - 9 = -2$
The equation is $y = (3/2)x - 2$

5. $3 = 100a + b$ or $b = 3 - 100a$
$12 = -60a + b$
Substituting into the second equation:
$12 = -60a + (3 - 100a)$
$12 = -160a + 3$
$-3 \quad -3$
$9 = -160a$
divide by -160
$a = 9/(-160) = -9/160$
Solving for b:
$b = 3 - 100(-9/160)$
$b = 69/5$
The equation is $y = (-9/160)x + 69/5$

7. $-4 = 7a + b$ or $-4 - 7a = b$
$5 = 0a + b$
Substituting into the second equation:
$5 = 0 + (-4 - 7a)$
$5 = -4 - 7a$
$+4 \quad +4$
$9 = -7a$
divide by -7
$a = -1/7$
Solving for b:
$b = -4 - 7(-9/7)$
$b = -4 + 9 = 5$
The equation is $y = (-9/7)x + 5$

9. $25 = 5000a + b$ or $b = 25 - 5000a$
$50 = 3000a + b$
Substituting into the second equation:
$50 = 3000a + 25 - 5000a$
$50 = -2000a + 25$
$-25 \quad -25$
$25 = -2000a$
divide by -2000
$a = 25/(-2000) = -1/80$
Solving for b:
$b = 25 - 5000(-1/80) = 87.5$
The equation is $y = (-1/80)x + 87.5$

11. Points on the line are (0, 5) and (20, 20)
$20 = 20a + b$ or $b = 20 - 20a$
$5 = 0a + b$
Substituting into the second equation:
$5 = 20 - 20a$
$-20 \quad -20$
$-15 = -20a$
divide by -20
$a = (-15)/(-20) = 3/4$
Solving for b:
$b = 20 - 20(3/4) = 20 - 15 = 5$
The equation is $y = (3/4)x + 5$

Homework

3. Using the regression capabilities of you calculator, the cost equation is: $y = -0.101x + 10.022$.

5. a. $7 = -2a + b$ or $b = 7 + 2a$
$-3 = 3a + b$
Substituting into the second equation:
$-3 = 3a + (7 + 2a)$
$-3 = 5a + 7$
$-7 \quad -7$
$-10 = 5a$
divide by 5
$a = -2$
Solving for b:
$b = 7 + 2\cdot(-2)$
$b = 3$
The equation is $y = -2x + 3$

b. $0 = (-2/3)a + b$ or $(2/3)a = b$
$3/4 = (5/2)a + b$
Substituting into the second equation:
$3/4 = (5/2)a + (2/3)a$
$3/4 = (19/6)a$
multiply by 6/19
$a = (6/19)\cdot(3/4)$
$a = 9/38$
Solving for b:
$b = (2/3)\cdot(9/38)$
$b = 3/19$
The equation is $y = (9/38)x + 3/19$

c. $2 = 1.5a + b$ or $2 - 1.5a = b$
$-6.9 = 5.125a + b$
Substituting into the second equation:
$-6.9 = 5.125a + (2 - 1.5a)$
$-6.9 = 3.625a + 2$
$-2 \quad -2$
$-8.9 = 3.625a$
divide by 3.625
$a = -2.455$
Solving for b:
$b = 2 - 1.5 \cdot (-2.455)$
$b = 5.683$
The equation is $y = -2.455x + 5.683$

d. $5 = 15a + b$ or $5 - 15a = b$
$20 = 50a + b$
Substituting into the second equation:
$20 = 50a + 5 - 15a$
$20 = 35a + 5$
$-5 \quad -5$
$15 = 35a$
divide by 35
$a = 15/35 = 3/7$
Solving for b:
$b = 5 - 15(3/7)$
$b = -10/7$
The equation is $y = (3/7)x - 10/7$

Lesson 4 The Harvest Model

Nitty Gritty: Equations from Graphs

1. $m = 3/3 = 1$
$b = 1$
$y = x + 1$

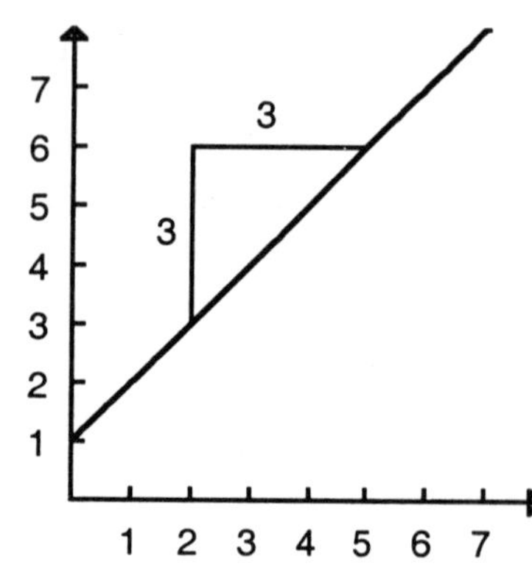

3. $m = 0/2 = 0$
$b = 4$
$y = 4$

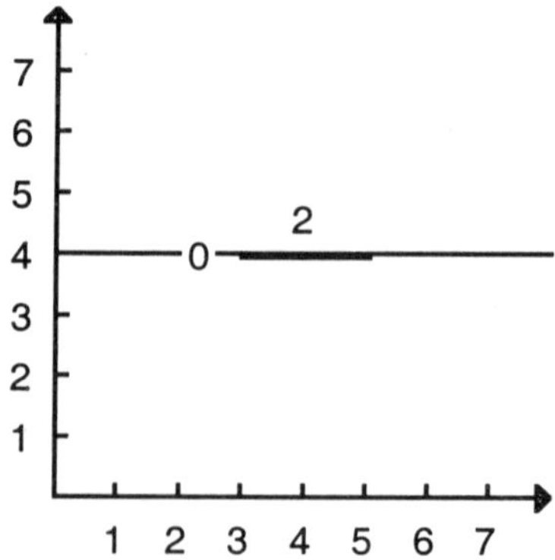

5. $m = -12$
   $b = 48$
   $y = -12x + 48$

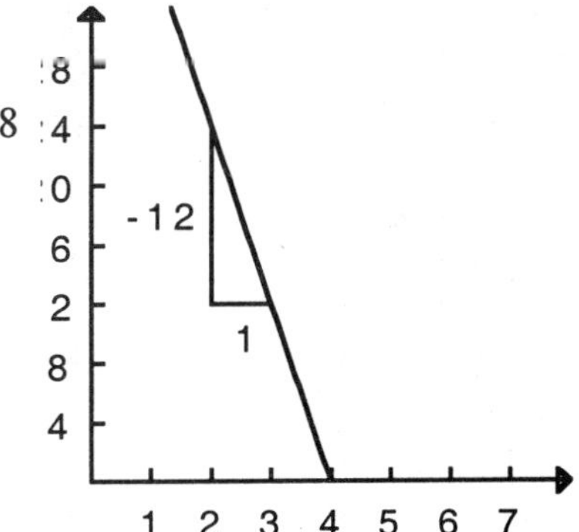

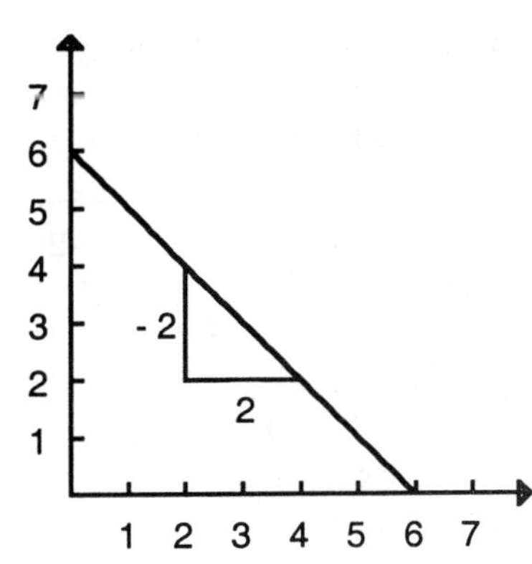

Homework

2. The fish must jump out of the water, wiggle up the beach, and hop into the frying pan for the cost per fish to equal 0.

6. $N = 12$ million fish
   $p = \$28$ dollars per fish
   $c = \$28$ dollars per fish

10. With a population of 20 million, the cost per fish is 8 shillings.
    Profit = Revenue – Cost
    $= 8.5 \cdot 25{,}000 - 8 \cdot 25{,}000$
    $= 212{,}500 - 200{,}000$
    $= 12{,}500$
    The profit for this catch is 2,500 shillings.

Lesson 5 The Business Model

Nitty Gritty: Solving Systems Using Substitution

1. $2x + y = 9$
   $3x - 4y = 8$

   $2x + y = 9$
   $-2x \quad -2x$
   $y = 9 - 2x$

   $3x - 4(9 - 2x) = 8$
   $3x - 36 + 8x = 8$
   $11x - 36 = 8$
   $+36 \quad +36$
   $11x = 44$
   divide both sides by 11
   $x = 4$

   $y = 9 - 2 \cdot 4$
   $y = 9 - 8 = 1$

   Solution: (4, 1)

3. $-2x + 7y = 10$
   $x - 3y = -3$

   $x - 3y = -3$
   $+3y \quad +3y$
   $x = 3y - 3$

   $-2(3y - 3) + 7y = 10$
   $-6y + 6 + 7y = 10$
   $y + 6 = 10$
   $-6 \quad -6$
   $y = 4$

   $x = 3 \cdot 4 - 3$
   $x = 12 - 3$
   $x = 9$

   Solution: (9, 4)

5. $5x + 16y = 15$
$-2x - 4y = 1$

$-2x - 4y = 1$
$+ 4y \quad +4y$
$-2x = 4y + 1$
divide by -2
$x = -2y - 1/2$

$5(-2y - 1/2) + 16y = 15$
$-10y - 5/2 + 16y = 15$
$6y - 5/2 = 15$
$+5/2 \quad +5/2$
$6y = 17.5$
divide by 6
$y = 2.9166667$

$x = -2\cdot(2.9166667) - 1/2$
$x = -6.33333$

Solution: $(-6.33333, 2.9166667)$

7. $2x + y = 3$
$8x - 4y = 8$

$2x + y = 3$
$-2x \quad -2x$
$y = -2x + 3$

$8x - 4(-2x + 3) = 8$
$8x + 8x - 12 = 8$
$16x - 12 = 8$
$+12 \quad +12$
$16x = 20$
divide by 16
$x = 1.25$

$y = -2\cdot 1.25 + 3$
$y = 0.5$

Solution: $(1.25, 0.5)$

9. $-9x + 5y = 1$
$4x - 2y = 2$

$4x - 2y = 2$
$-4x \quad -4x$
$-2y = -4x + 2$
divide by -2
$y = 2x - 1$

$-9x + 5(2x - 1) = 1$
$-9x + 10x - 5 = 1$
$x - 5 = 1$
$+5 \quad +5$
$x = 6$

$y = 2\cdot 6 - 1$
$y = 12 - 1$
$y = 11$

Solution: $(5, 11)$

11. $-2x + y = 1$
$x - 3y = 2$

$-2x + y = 1$
$+2x \quad +2x$
$y = 2x + 1$

$x - 3(2x + 1) = 2$
$x - 6x - 3 = 2$
$-5x - 3 = 2$
$+3 \quad +3$
$-5x = 5$
divide by -5
$x = -1$

$y = 2\cdot(-1) + 1$
$y = -2 + 1$
$y = -1$

Solution: $(-1, -1)$

13. $3.5x + 0.25y = -2$
$0.3x - 0.1y = 2.25$

$3.5x + 0.25y = -2$
$-3.5x \quad -3.5x$
$0.25y = -3.5x - 2$
divide by 0.25
$y = -14x - 8$

$0.3x - 0.1(-14x - 8) = 2.25$
$0.3x + 1.4x + 0.8 = 2.25$
$1.7x + 0.8 = 2.25$
$-0.8 \quad -0.8$
$1.7x = 1.45$
divide by 1.7
$x = 0.853$

$y = -14\cdot(0.853) - 8$
$y = -11.942 - 8$
$y = -19.942$

Solution: $(0.853, -19.942)$

15. $1.25x + 2.5y = -3$
$-0.5x - 1.2y = 1.5$

$1.25x + 2.5y = -3$
$-2.5y \quad -2.5y$
$1.25x = -2.5y - 3$
divide by 1.25
$x = -2y - 2.4$

$-0.5(-2y - 2.4) - 1.2y = 1.5$
$y + 1.2 - 1.2y = 1.5$
$-0.2y + 1.2 = 1.5$
$-1.2 \quad -1.2$
$-0.2y = 0.3$
divide by -0.2
Solution: $(-0.469, -5.469)$
$y = -1.5$

$x = -2\cdot(-1.5) - 2.4$
$x = 3 - 2.4$
$x = 0.6$

Solution: $(0.6, -1.5)$

Homework

1. Revenue = (market price)·(quantity)
   = (8.5)·(25,000)
   = 212,500 shillings

   Costs = fixed costs + variable costs
   = 60,000 + (cost per fish)·(quantity)
   = 60,000 + (6)·(25,000)
   = 60,000 + 150,000
   = 210,000 shillings

   Profit = Revenue – Cost
   = 212,500 – 210,000
   = 2,500 shillings

4. Answers vary. One concept map might be:

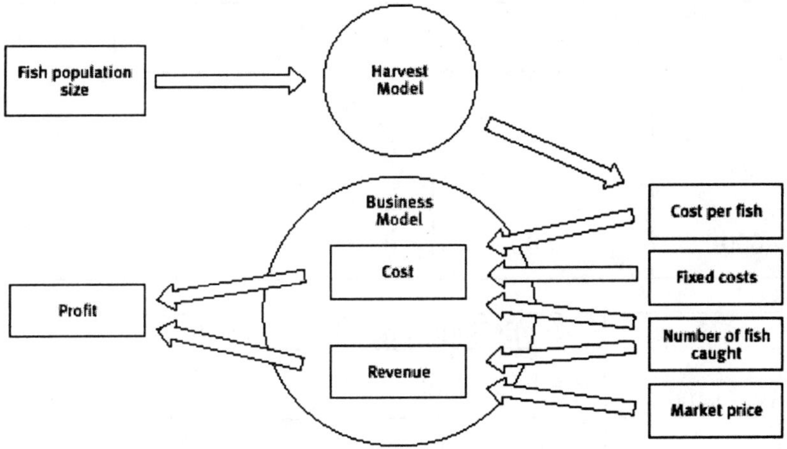

8. No, the cost does not double when the time doubles.

13. a.

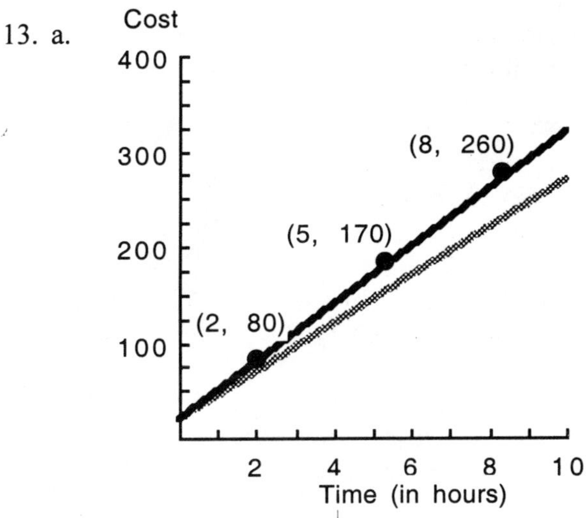

   b. The new line is steeper.

c. The vertical intercept, C-intercept, is the same.

d. The new slope, m, is 30, the intercept, b, is 20. Equation: $C = 30t + 20$.

15. a. The line for electrician A is steeper, thus he has a higher per-hour fee.

b. The line for electrician B has a higher C-intercept, thus he has a higher truck charge.

c. It depends on how long the job takes. If you think the job will take less than 5 hours, hire electrician A. If you think the job will take more than 5 hours, hire electrician B.

Lesson 6 Big Net

Homework

1. a.

	Yearly Catch Size	Market Price	Fixed Costs	Cost per Fish
Last Year (small net)	7,000	$4.00	$6,400	$2.50
Current Year (big net)	14,000	$4.00	$14,400	$3.25
Next Year (big net)	14,000	$4.00	$6,400	$3.25

	Business Model	Profit Equation	Profit or Loss
Last Year (small net)	$R = 4x$ $C = 2.5x + 6400$	$P = 1.5x - 6400$	$P = 1.5\cdot 7{,}000 - 6{,}400$ $= \$4{,}100$
Current Year (big net)	$R = 4x$ $C = 3.25x + 14{,}400$	$P = 0.75x - 14{,}400$	$P = 0.75\cdot 14{,}000 - 14{,}400$ $= -\$3{,}900$
Next Year (big net)	$R = 4x$ $C = 3.25x + 6{,}400$	$P = 0.75x - 6{,}400$	$P = 0.75\cdot 14{,}000 - 6{,}400$ $= \$4{,}100$

b. Ahab should not purchase the larger net under these circumstances. He is not improving his long-term profit if the cost per fish increases to $3.25 while the market price stays $4.00.

Lesson 7 The Market Model

Nitty Gritty: Solving Systems Using Elimination

1. $3x + 2y = 7$
   $x + 3y = 14$
   multiply (2) by –3

   $3x + 2y = 7$
   $\underline{-3x - 9y = -42}$
   $-7y = -35$
   divide by –7
   $y = 5$

   $3x + 2y = 7$
   $3x + 2 \cdot 5 = 7$
   $3x + 10 = 7$
   $-10 \quad -10$
   $3x = -3$
   divide both sides by 3
   $x = -1$
   Solution: (–1, 5)

3. $8x - 2y = -14$
   $12x + 3y = 9$
   multiply (1) by 3
   multiply (2) by 2

   $24x - 6y = -42$
   $\underline{24x + 6y = 18}$
   $48x = -24$
   divide by 48
   $x = -1/2$

   $8x - 2y = -14$
   $8(-1/2) - 2y = -14$
   $-4 - 2y = -14$
   $+4 \quad +4$
   $-2y = -10$
   divide both sides by –2
   $y = 5$
   Solution: (–1/2, 5)

5. $12x - 4y = 3$
   $17x - 5y = 8$
   multiply (1) by 5
   multiply (2) by –4

   $60x - 20y = 15$
   $\underline{-68x + 20y = -32}$
   $-8x = -17$
   divide by –8
   $x = 2.125$

   $12x - 4y = 3$
   $12(2.125) - 4y = 3$
   $25.5 - 4y = 3$
   $-25.5 \quad -25.5$
   $-4y = -22.5$
   divide by –4
   $y = 5.625$
   Solution: (2.125, 5.625)

7. $8x - 2y = 38$
   $12x + 3y = 39$
   multiply (1) by 3
   multiply (2) by 2

   $24x - 6y = 114$
   $\underline{24x + 6y = 78}$
   $48x = 192$
   divide by 48
   $x = 4$

   $8x - 2y = 38$
   $8(4) - 2y = 38$
   $32 - 2y = 38$
   $-32 \quad -32$
   $-2y = 6$
   divide by –2
   $y = -3$
   Solution: (4, –3)

9. $-9x + 5y - 1 = 0$
   $4x - 2y - 2 = 0$
   multiply (1) by 2
   multiply (2) by 5

   $-18x + 10y - 2 = 0$
   $\underline{20x - 10y - 10 = 0}$
   $2x - 12 = 0$
   $+12 \quad +12$
   $2x = 12$
   divide by 2
   $x = 6$

   $4x - 2y - 2 = 0$
   $4(6) - 2y - 2 = 01$
   $24 - 2y - 2 = 0$
   $22 - 2y = 0$
   $+2y \quad +2y$
   $22 = 2y$
   divide by 2
   $y = 11$
   Solution: (6, 11)

11. $3x + 2y = 6$
    $-5x - 4y = -8$
    multiply (1) by 2

    $6x + 4y = 12$
    $\underline{-5x - 4y = -8}$
    $x = 4$

    $3x + 2y = 6$
    $3(4) + 2y = 6$
    $12 + 2y = 6$
    $-12 \quad -12$
    $2y = -6$
    divide by 2
    $y = -3$
    Solution: (4, –3)

13. $5x + 7y + 1 = 0$
$2x = 3y$
multiply (1) by 2
multiply (2) by –5

$10x + 14y + 2 = 0$
$-10x = -15y$
$14y + 2 = -15y$
$-14y \quad -14y$
$2 = -29y$
divide by –29
$y \approx -0.069$

$2x = 3y$
$2x = 3(-0.069)$
$2x = -0.207$
divide by 2
$x \approx -0.104$
Solution: (–0.104, –0.069)

15. $7x - 2y = 41$
$-2x - y = -7$
multiply (2) by –2

$7x - 2y = 41$
$4x + 2y = 14$
$11x = 55$
divide by 11
$x = 5$

$7x - 2y = 41$
$7(5) - 2y = 41$
$35 - 2y = 41$
$-35 \quad -35$
$-2y = 6$
divide by –2
$y = -3$
Solution: (5, –3)

Homework

3. a. supply: $p = 0.2 \cdot N + 2$
demand: $p = -0.7 \cdot N + 20$
By substitution: $-0.7 \cdot N + 20 = 0.2 \cdot N + 2$
$-0.2N \quad -0.2N$
$-0.9N + 20 = 2$
$-20 \quad -20$
$-0.9N = -18$
divide both sides by –0.9
$N = 20$

$p = 0.2 \cdot 20 + 2$
$p = 4 + 2 = 6$
The solution is (20, 6). The solution reveals that 6 shillings is the selling price at equilibrium, and 20 million fish should be sold.

b. 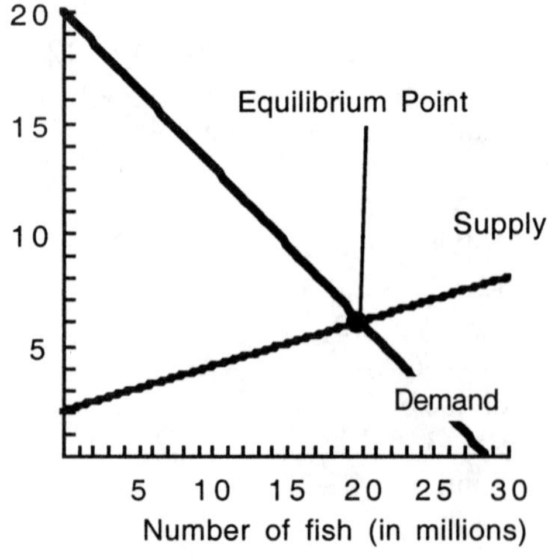

c. This year's market price: 6 shillings. This year's market size: 20 million fish. This year's total catch size: 20 million fish.

d. Revenue = (market price)·(catch size)·(market share)
= (6)(20,000,000)(0.001)
= 120,000 shillings

Cost = fixed cost + (cost per fish)·(catch size)·(market share)
= 50,000 + (3)(20,000,000)(0.001)
= 110,000 shillings

Profit = Revenue – Cost
= 120,000 – 110,000
= 10,000 shillings

4. a. supply: $p = 0.2 \cdot N + 1$
demand: $p = -0.6 \cdot N + 15$
By substitution:
$$0.2 \cdot N + 1 = -0.6 \cdot N + 15$$
$$+0.6N \qquad +0.6N$$
$$0.8N + 1 = 15$$
$$-1 \quad -1$$
$$0.8N = 14$$
divide both sides by 0.8
$$N = 17.5$$
$p = 0.2 \cdot 17.5 + 1$
$p = 3.5 + 1 = 4.5$
The solution is (17.5, 4.5). The solution reveals that 4.5 shillings is the selling price at equilibrium, and 17.5 million fish should be sold.

b. 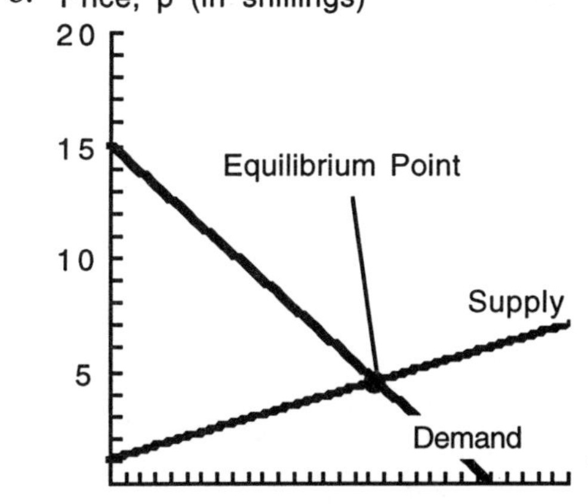

c. This year's market price: 4.5 shillings. This year's market size: 17.5 million fish. This year's total catch size: 17.5 million fish.

d. Revenue = (market price)·(catch size)·(market share)
= (4.5)(20,000,000)(0.001)
= 78,750 shillings

Cost = fixed cost + (cost per fish)·(catch size)·(market share)
= 50,000 + (2.5)(20,000,000)(0.001)
= 93,950 shillings

Profit = Revenue – Cost
= 78,750 – 93,950
= –15,200 shillings

7.

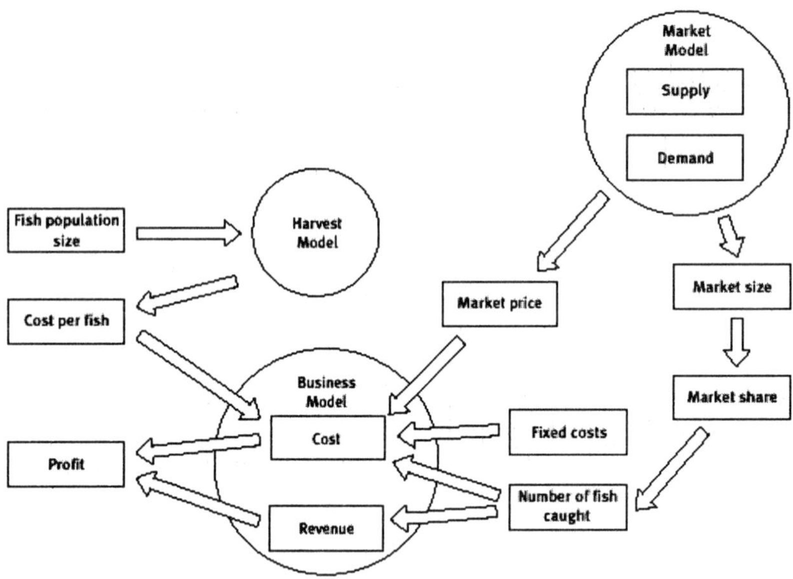

10. a. $-x + 2y = 2$
$\underline{x - y = 5}$
$y = 7$

$x - y = 5$
$x - 7 = 5$
$+7 \quad +7$
$x = 12$
Solution: (12, 7)

b. $x - 4y = 3$
$\underline{5x + 4y = 9}$
$6x = 12$
divide by 6
$x = 2$

$x - 4y = 3$
$2 - 4y = 3$
$-2 \qquad -2$
$-4y = 1$
divide by –4
$y = -1/4$
Solution: (2, –1/4)

c. $3x + 4y = 7$
$2x - 5y = -3$
multiply (1) by 5
multiply (2) by 4

$15x + 20y = 35$
$8x - 20y = -12$
$23x = 23$
divide by 23
$x = 1$

$3x + 4y = 7$
$3(1) + 4y = 7$
$3 + 4y = 7$
$-3 \quad -3$
$4y = 4$
divide by 4
$y = 1$
Solution: (1, 1)

d. $3x - 2y = 8$
$x + 4y = 5$
multiply (1) by 2

$6x - 4y = 16$
$x + 4y = 5$
$7x = 21$
divide by 7
$x = 3$

$x + 4y = 5$
$3 + 4y = 5$
$-3 \quad -3$
$4y = 2$
divide by 4
$y = 1/2$
Solution: (3, 1/2)

Lesson 8 Harvesting and the Growth Model

Homework

2. If they did overlap, significant numbers of female fish would be caught before they could lay their eggs. This would noticeably affect the births per year and, therefore, the growth per year curve.

7. a. A few fish with an abundance of food means the fish population size increases rapidly (that is, there is a large percent growth). The fish are very healthy and fertile. Lots of fish with the same amount of food means the fish population size still increases, but not as fast as (that is, the percent growth is smaller). The fish are less healthy and fertile. Too many fish for the same amount of food as earlier means the fish population begins to starve and the fish population size begins to shrink (that is, the percent growth is negative). Therefore, percent growth should decrease as fish population size increases.

   b. The further below the carrying capacity the fish population is maintained, the larger the population buffer for natural disasters. That is, a natural disaster can significantly lower the carrying capacity without forcing a large drop in fish population, if the fish population size was maintained significantly below the original carrying capacity.

8. a. Using figure 17, if the current fish population size is 10,000, the growth is approximately 7,000 fish.

   b. The population is 10,000 + 7,000 = 17,000 fish. If 9,000 fish are caught, as predicted, next year's population size will be 8,000 fish.

   c. With a population of 8,000 fish, the cost per fish is approximately $9.20.

Lesson 9 Implications for Public Policy

Homework

1. a. Revenue = (market price)(market share)(market size)
$R = (13)(0.001)(16{,}000{,}000)$
$R = 208{,}000$ shillings
Cost = fixed costs + (cost per fish)(market share)(total catch size)
$C = 60{,}000 + (7)(0.001)(16{,}000{,}000)$
$C = 172{,}000$ shillings
Profit = Revenue – Cost
$P = 208{,}000 - 172{,}000$
$P = 36{,}000$ shillings

b. According to the growth model, with a population of 30,000,000, the growth will be 25,000,000. Next year's population will be $30 + 25 - 16 = 39$ million.

c. According to the harvest model, with a population of 39,000,000 the cost per fish is 6.1 shillings.
$R = (13)(0.001)(16{,}000{,}000) = 208{,}000$ shillings
$C = 60{,}000 + (6.1)(0.001)(16{,}000{,}000) = 97{,}600$ shillings
$P = 208{,}000 - 97{,}600 = 50{,}400$ shillings

2. a. $R = (16.25)(0.001)(14{,}000{,}000) = 227{,}500$ shillings
$C = 60{,}000 + (7)(0.001)(14{,}000{,}000) = 158{,}000$ shillings
$P = 227{,}500 - 158{,}000 = 69{,}500$ shillings

b. According to the growth model, with a population of 30,000,000 the growth will be 25,000,000. The new fish population is $30 + 25 - 14 = 41$ million fish.

c. According to the harvest model, with a population of 41,000,000 the cost per fish is 5.9 shillings.
$R = (16.25)(0.001)(14{,}000{,}000) = 227{,}500$ shillings
$C = 60{,}000 + (5.9)(0.001)(14{,}000{,}000) = 142{,}600$ shillings
$P = 227{,}500 - 142{,}600 = 84{,}900$ shillings

6. a. $(3x + 4)(2x - 1)$
$= 6x^2 - 3x + 8x - 4$
$= 6x^2 + 5x - 4$

b. $(2x - 5)(x + 8)$
$= 2x^2 + 16x - 5x - 40$
$= 2x^2 + 11x - 40$

c. $(7.5x + 21)(3x - 11.2)$
$= 22.5x^2 - 84x + 63x - 235.2$
$= 22.5x^2 - 21x - 235.2$

d. $(-2.3x + 7)(3x - 5)$
$= -6.9x^2 + 11.5x + 21x - 35$
$= -6.9x^2 + 32.5x - 35$

e. $(x^2 + 2x - 3)(x + 5)$
$= x^2(x + 5) + 2x(x + 5) - 3(x + 5)$
$= x^3 + 5x^2 + 2x^2 + 10x - 3x - 15$
$= x^3 + 7x^2 + 7x - 15$

f. $(2x^2 - 3x)(3x - 8)$
$= 6x^3 - 16x^2 - 9x^2 + 24x$
$= 6x^3 - 25x^2 + 24x$

Lesson 10 Balancing Economic and Ecological Interests

Homework

3.

Fish population size (1000's of fish)	Growth per year (1000's of fish) (1)	Market price (dollars) (2)	Revenue (1000's of dollars) (1)•(2) = (3)	Cost per fish (dollars) (4)	Cost (1000's of dollars) (1)•(4) = (5)	Profit (1000's of dollars) (3) – (5)
0	0.0	24	0	55.0	0.0	0.0
5	8.75	24	210	40.35	353.1	–143.1
10	15.0	24	360	30.0	450.0	–90.0
15	18.75	24	450	22.67	425.1	24.9
20	20.0	24	480	17.50	350.0	130.0
25	18.75	24	450	13.84	259.5	190.5
30	15.0	24	360	11.25	168.8	191.3
35	8.75	24	210	9.41	82.3	127.7
40	0.0	24	0	8.13	0.0	0.0

6. From the graph, the break-even points is approximately (14,000, 436,000).

10. $C = 49.72 \cdot (0.9529)^N$

11. $y = -0.05N^2 + 2N$

Calculator Keystroke Manual

TI-82

TI-83

TI-83Plus

TI-85

TI-86

To accompany the

Maricopa Mathematics Modules

Preliminary Edition

Calculator Keystroke Manual for the TI-82

Contents

Introduction

The graphing calculator is a tool to enable students to visualize mathematics, solve problems, think critically, and develop a conceptual understanding of the mathematics being studied. This Calculator Keystroke Manual is to provide students with a reference manual for using a graphing calculator. The intent is to allow students to focus on learning mathematics and for teachers to teach mathematics rather than the focus being on pushing buttons. It is written so that it can be referenced whenever a particular calculator function is desired. Each function, listed in alphabetical order, contains a step-by-step approach with calculator screen shots displaying what should be seen after each step. Occasionally, sample problems are given and worked through to better display that calculator function. Anytime a particular calculator button is referenced, the button name will be typed in all CAPS. If a 2nd function is referenced, the main button name is used with the actual function desired typed in parenthesis. For example, to turn the calculator OFF, type 2nd ON (OFF).

This manual does not provide instructions on using every function available on the calculator. With this background and subsequent mathematics courses, students will continue to develop their ability to use the calculator as an effective and efficient tool for doing mathematics.

Before beginning, be sure that the calculator is prepared correctly.

Preparing the Calculator

1. Press MODE. For the purposes of this manual, all settings on the left hand side of the screen should be highlighted. If something is not set properly, use the UP, DOWN, LEFT and RIGHT keys to move the cursor to the desired location. Press ENTER to highlight the appropriate setting.

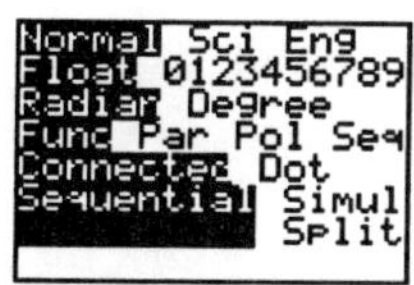

2. Press 2nd $Y=$ (STAT PLOT). Use the DOWN arrow key to move the cursor to option 4:PlotsOff. Press ENTER.

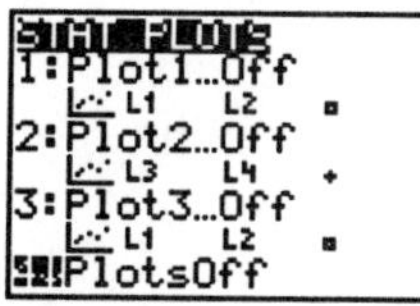

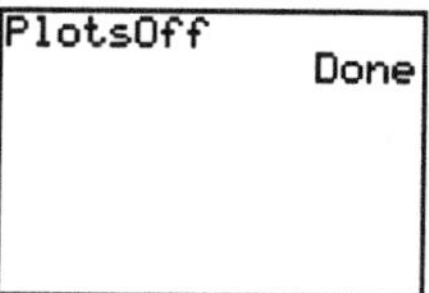

3. The HOME SCREEN is the main screen where most calculations are done.

Using This Manual

A good way to use this manual is to complete each of sections in the order given, then use the manual as a reference when needed.

1. ON/OFF
2. Contrast
3. ANS (Answer)
4. DEL/INS (Delete/Insert)
5. ENTRY
6. Square Root
7. Exponents
8. Scientific Notation
9. QUIT
10. Fractions
11. $Y=$
12. ZOOM/WINDOW
13. TRACE
14. Tables
15. Absolute Value
16. Value
17. Evaluate a Function
18. Intersect
19. Root
20. One-Variable Statistics
21. Histograms
22. Box and Whisker Plots
23. Regression/Scatter Plots/Lists
24. Error Messages
25. Trigonometric Functions

Absolute Value

The absolute value of a number always returns the positive value of that number. The absolute value function can be performed numerically in the home screen or can be graphed as a function in the $y =$ screen.

Example: Determine the absolute value of –5.

1. From the home screen, press 2nd x^{-1} (ABS).

```
abs
```

2. Enter –5 in the parentheses. Be sure to use the gray negative button not the blue minus button. Press ENTER.

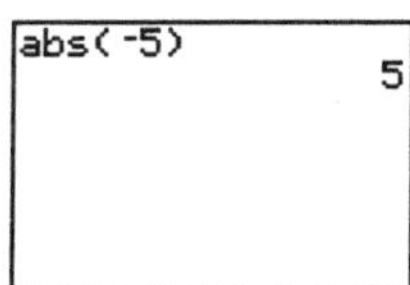

Example: Graph $y_1 = |x|$.

3. Press y = and enter y_1 = abs(x). Follow the previous instructions for accessing the abs command.

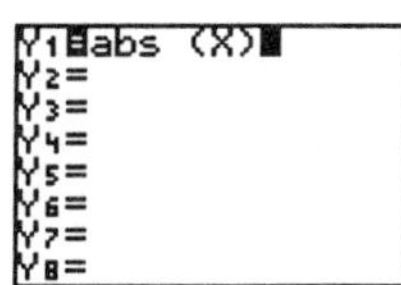

4. Press ZOOM 6: ZStandard to graph the function in the standard viewing window.

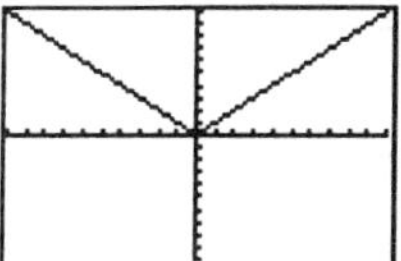

ANS

The ANS function recalls the previous answer so that additional operations can be performed on the answer to a previous operation.

Example: Start with the number 1 and continue to double it indefinitely.

1. Enter the number 1 and press ENTER. Press the multiplication button (×) then press 2 and ENTER. The Ans part of the display refers to the previous answer, in this case the 1 that was entered initially.

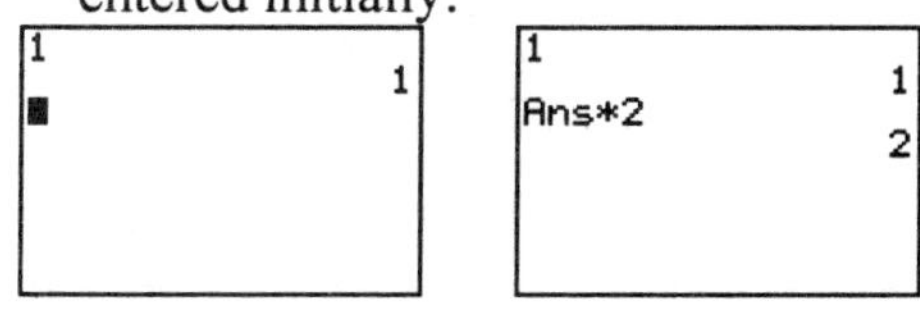

2. Continue to press ENTER until reaching the answer 64. Each time, the calculator takes the previous answer and multiples by 2.

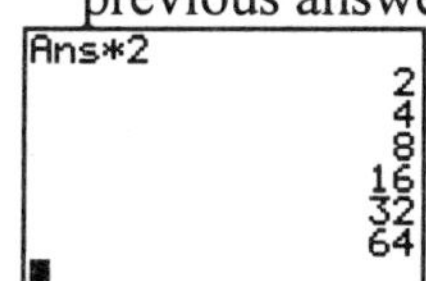

The Ans function can be accessed by typing 2nd (-) (ANS).

3. Press 7, press +, press 2nd (-) (ANS), then ENTER. The calculator takes the last answer (64), adds 7, and returns the answer of 71.

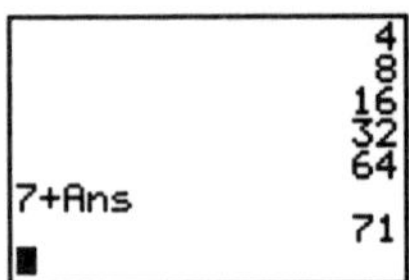

Box and Whisker Plots

Box and whisker plots are used to display and compare data. Create a box and whisker plot for the data given.

The ACT scores for some students in a particular school is given.

Student	1	2	3	4	5	6	7	8	9
ACT Score	20	22	30	32	28	21	20	34	18

1. Press STAT and select 1: EDIT.

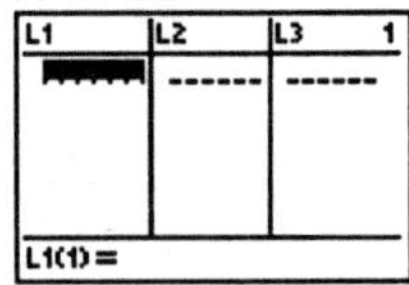

This will take you to the TI version of a spreadsheet called Lists. If data already is entered into the lists, you can clear it by using the UP ARROW to highlight the L1 at the very top of the screen. When highlighted, press CLEAR and ENTER. Repeat for L2, etc.

Enter the data into the lists by typing in the appropriate numbers and pressing ENTER after each entry. For this example, enter only the ACT scores into L1.

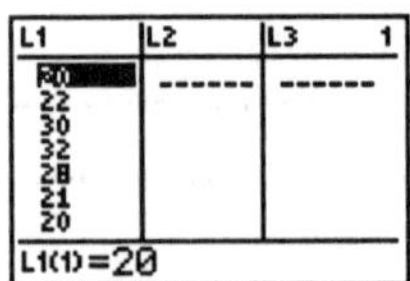

2. Set up the plot by pressing 2nd y = (STAT PLOT). Choose Plot 1, 2, or 3. Make sure that the plot is turned ON, the type is Box and Whisker (2nd to the last choice), the Xlist is the list containing the data (L_1 in this case), and the frequency (Freq) is 1.

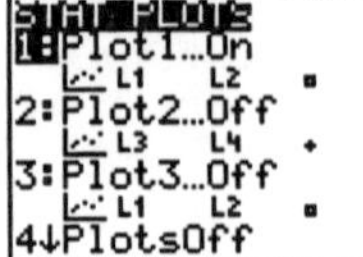

3. Adjust your WINDOW so that it reflects the values entered into the lists. For a Box and Whisker plot, only the Xmin and Xmax must reflect the data. The Ymin and Ymax is irrelevant.

WINDOW FORMAT
Xmin=0
Xmax=36
Xscl=1
Ymin=0
Ymax=.5
Yscl=1

4. Use the TRACE command and your RIGHT and LEFT ARROW keys to find the minimum data value, first quartile, median, third quartile, and maximum data value.

minX=18 Q1=20 Med=22 Q3=31 maxX=34

Contrast

The contrast for the calculator screen can be adjusted to be darker or lighter. As your batteries become older, it will become necessary to adjust the contrast so that you can more easily read the calculator screen.

1. Press 2nd and release.

2. Hold the DOWN ARROW key to make the screen lighter. Hold the UP ARROW key to make the screen darker. Each time the DOWN or UP ARROW key is released, the 2nd key must be pressed again to continue to adjust the contrast.

3. As the UP or DOWN ARROW keys are held, notice the number in the upper right hand corner. The contrast is measured from 1 – 9 where 1 is the lightest setting and 9 is the darkest. If the contrast needs to be 9 in order for you to see the screen, you probably need to replace the batteries.

DEL/INS

The DEL (Delete) key deletes the character under the current cursor location. The 2nd DEL (INS) (Insert) key inserts a character to the left of the current cursor location.

Example: You desire to type 9+6-5*34, but accidentally type 9+*36*-5**4*.

1. Enter 9+*36*-5**4*. Use the LEFT arrow to move the cursor to the 3 location.

9+36-5*4

2. Press DEL. This deletes the unneeded 3.

3. Use the RIGHT ARROW to move the cursor to the 4 location. Press 2nd DEL (INS) and enter a 3. This inserts the 3 to the left of the 4 as desired. Notice that the cursor changed after pressing 2nd DEL (INS) so that you knew that the calculator was in the insert mode.

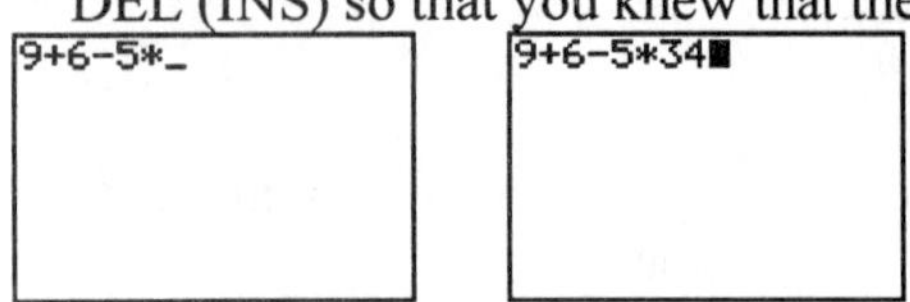

ENTRY

The ENTRY function returns previously entered operations on the calculator screen so that you may edit them.

1. Press 2nd ENTER (ENTRY). The last operation entered into the computer appears on the screen and the cursor is at the right end of the line. Use the UP, DOWN, RIGHT, LEFT arrow keys to edit the line.

2. Press 2nd ENTER (ENTRY) again. Continue to press 2nd ENTER (ENTRY) and notice that previous entries continue to be restored on the screen. Any one of them can be edited.

Error Messages

Occasionally, error messages are given to indicate some problem with the function you are asking the calculator to perform. Three common error messages and how to correct them are provided.

Syntax Error
Syntax refers to the way in which the function or command was entered. One common error deals with use of parentheses.

Example: You intend to enter 6*(3+2) but accidentally enter 6*3+2).

1. In the Home Screen, enter 6*3+2) and press ENTER.

6*3+2)

ERR:SYNTAX
1:Quit
2:Goto

2. Because of the missing opening parenthesis, the calculator returns a syntax error. Notice the options on the screen. You can QUIT and start over, or you can GOTO the error so that it can be corrected. Select GOTO by pressing ENTER or the number 2.

3. To correct the mistake, insert a parenthesis before the 3 using INS. Press ENTER.

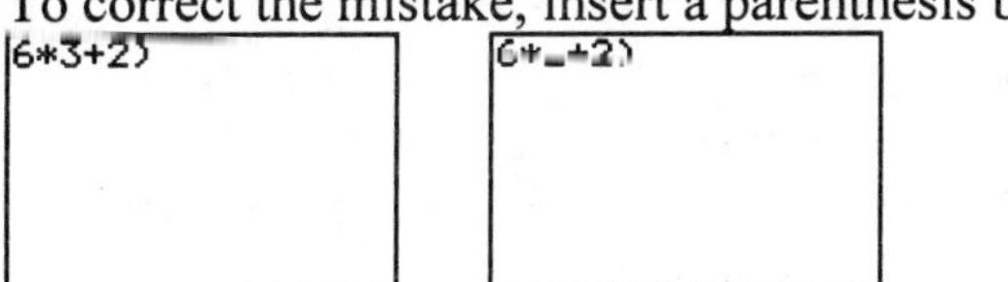

Window Range Error

The Graph Window Error refers to an incorrect WINDOW entry.

1. Enter an equation into $Y =$.

For example: $Y_1 = x^3 + 3$

2. Press ZOOM 6: ZStandard to graph the function in the standard viewing window.

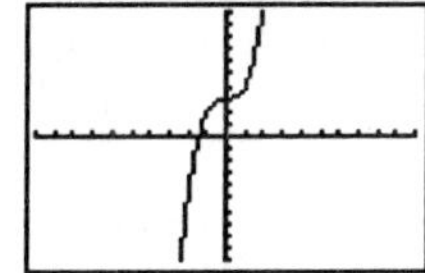

3. Press WINDOW. Imagine that you incorrectly changed the WINDOW as shown below. Press GRAPH. A WINDOW RANGE Error occurs.

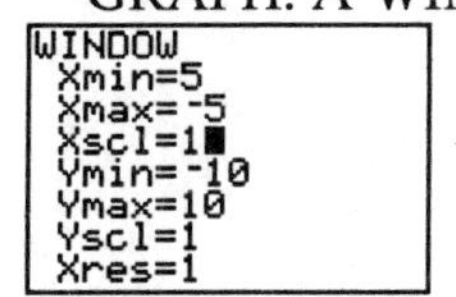

4. The only option is to QUIT. To correct the error, press WINDOW and change the Xmin and Xmax as shown. Press GRAPH.

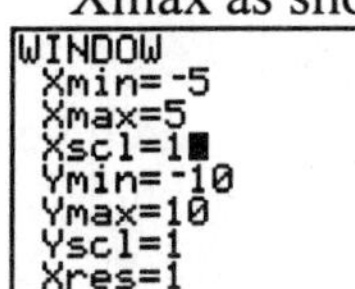

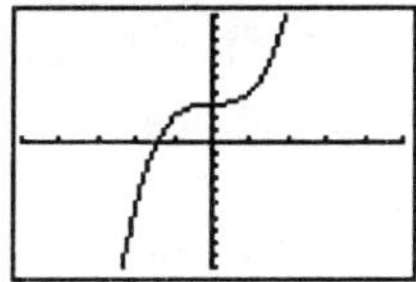

Dimension Mismatch Error (DIM MISMATCH)

The Dimension Mismatch error results when the data entered in the Lists do not match up. That is, there are more data entered in one list than in the other.

1. When DIM MISMATCH is encountered, the only option is to QUIT (option 1).

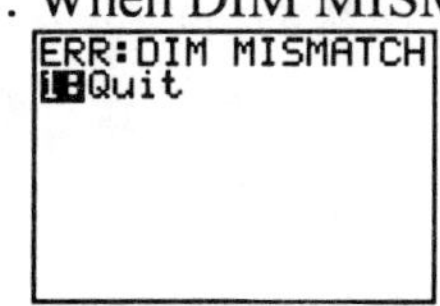

2. Press QUIT. To correct the error, refer to Preparing the Calculator, step 2.

Evaluate a Function

One way to evaluate a function is to use the notation $Y_1(x)$ in the home screen.

1. Enter an equation into Y_1.

Example: $Y_1 = 2x^2 + 3x + 1$

2. Press 2nd MODE (QUIT) to return to the home screen.

3. Press 2nd VARS (*Y*-VARS), and select option 1:Function.

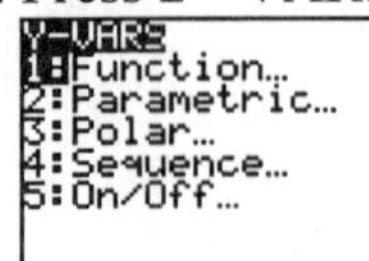

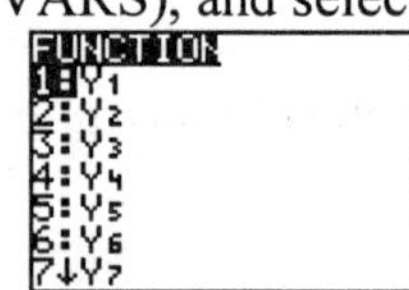

4. Since the equation was entered in Y_1, select option 1: Y_1. Type a left parenthesis, enter a value for x, close the parenthesis and press ENTER.

Example: $Y_1(2)$

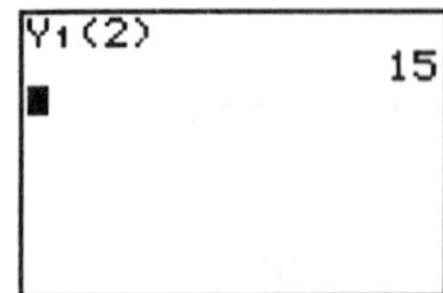

5. If many evaluations need to be completed, press 2nd ENTER (ENTRY) to repeat the previous entry, then use arrows to move the cursor over the value of x that needs to be changed and type in a new value for x and press ENTER.

Exponents

Exponents of power 2 can be entered into the calculator 2 different ways. Exponents of 2 or greater are entered using the caret key.

Example: Evaluate $3^2 + 7^2 - 8^4 + 2^7$.

1. To enter the powers of 2, use the x^2 key that is located in the middle of the first column of keys. To enter powers of 2 or greater, use the caret key (∧) which is located below the CLEAR key in the last column.

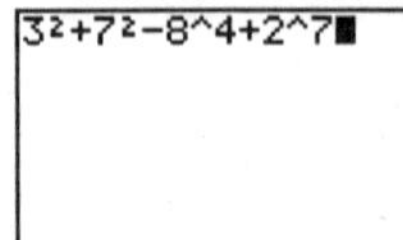

2. Press ENTER.

```
3²+7²-8^4+2^7
            -3910
```

Fractions

The TI-82 can convert from fractions to decimals and from decimals to fractions.

Example: Add $\frac{3}{8}+\frac{5}{12}$. Write your answer as a fraction.

1. Enter 3/8 + 5/12 and press ENTER. The fraction bar (/) is the division symbol on the keypad.

```
3/8+5/12
     .7916666667
```

2. To change the answer from a decimal to a fraction, press MATH and select option 1: ▶ Frac.

```
3/8+5/12
     .7916666667
Ans▸Frac
           19/24
```

Note: See the ANS section if you don't know what Ans means.

3. To convert the fraction back to a decimal, press MATH and select option 2: ▶ Dec.

```
3/8+5/12
     .7916666667
Ans▸Frac
           19/24
Ans▸Dec
     .7916666667
```

This screen shows that the fraction $\frac{19}{24}$ is equivalent to the decimal 0.7916666667.

Histograms

Histograms are a way to display and analyze data. Create a histogram for the data given.

The shoe size of some the children in the class is shown.

Child	1	2	3	4	5	6	7	8	9
Shoe Size	5	5	4	5	4	3	6	5	4

1. Press STAT and select 1: EDIT.

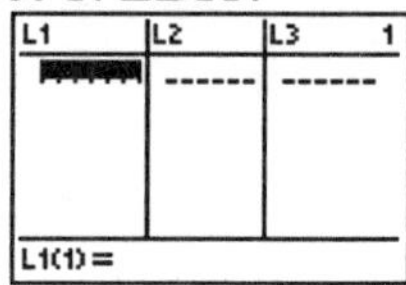

This will take you to the TI version of a spreadsheet called Lists. If data already is entered into the lists, you can clear it by using the UP ARROW to highlight the L1 at the very top of the screen. When highlighted, press CLEAR and ENTER. Repeat for L2, etc.

Enter the data into the lists by typing in the appropriate numbers and pressing ENTER after each entry. For this example, enter only the shoe sizes into L1.

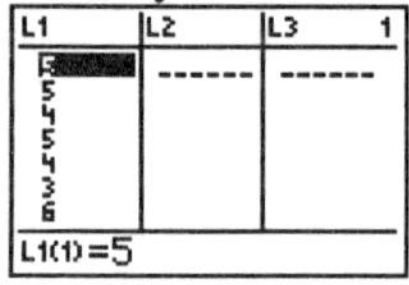

2. Set up the plot by pressing 2nd $Y =$ (STAT PLOT). Choose Plot 1, 2, or 3. Make sure that the plot is turned ON, the type is Histogram (4th choice), the Xlist is the list containing the data (L_1 in this case), and the frequency (Freq) is 1.

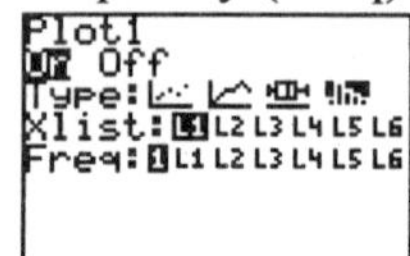

3. Adjust your WINDOW so that it reflects the values entered into the lists. The Ymin could be set at 0 because 0 is the least number of times a particular shoe size is seen. Ymax could be set at 5 since the largest number of times a particular shoe size is seen is 4 (4 children wear size 5 shoes).

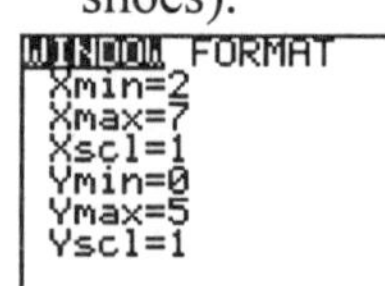

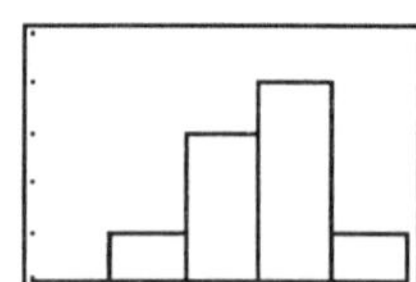

4. Use the TRACE command and your RIGHT and LEFT ARROW keys to find the frequency of each shoe size.

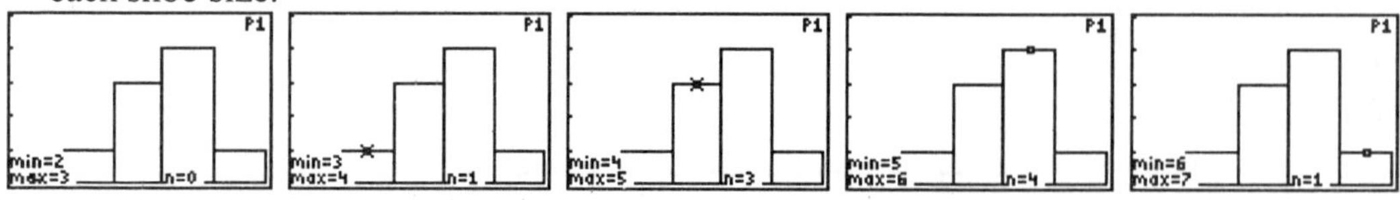

Intersect

The intersect function will calculate an intersection point of two graphed equations.

1. Enter the equations in Y_1 and Y_2.

   Example: $Y_1 = 2x^2 + 3x + 1$

   $Y_2 = 3x + 4$

2. Select an appropriate viewing window. In this case the standard viewing window may be used (ZOOM 6).

3. Either press GRAPH or ZOOM 6.

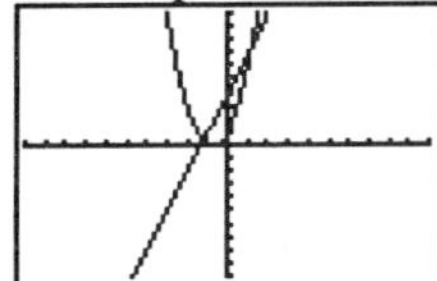

4. To determine the first quadrant point of intersection for these two graphs, press 2nd TRACE (CALC) and select option 5 (intersect).

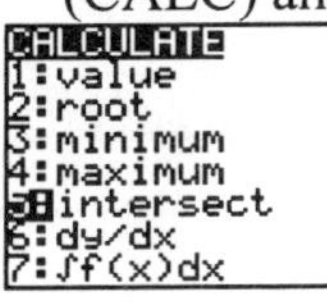

5. Automatically returning to the graph screen, the calculator asks for the First curve. The First curve is the function defined in Y_1. In the upper right hand corner of the screen, a 1 should be showing, representing that the equation entered into Y_1 is being selected. If it is, press ENTER. If not, press either the UP or DOWN arrow to switch to that equation, then press ENTER.

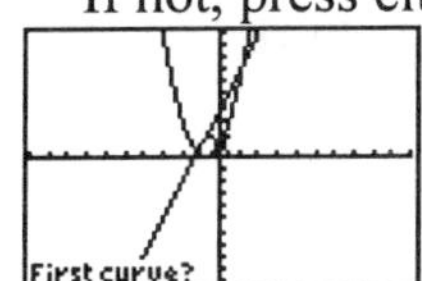

6. The calculator now asks for the Second curve. The Second curve is the function defined in Y_2 In the upper right hand corner of the screen, a 2 should be showing, representing that the equation entered into Y_2 is being selected. If it is, press ENTER. If not, press either the UP or DOWN arrow to switch to that equation, then press ENTER.

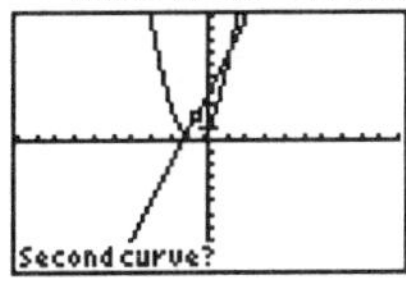

7. The calculator next asks for a guess as to where the point of intersection you are looking for is located on the graph. Use the LEFT and RIGHT arrow keys to move the cursor to the place where the graphs intersect in the first quadrant (in this example).

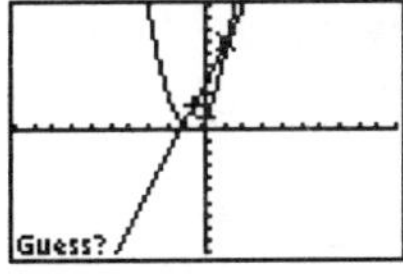

8. The calculator will show the point of intersection at the bottom of the screen.

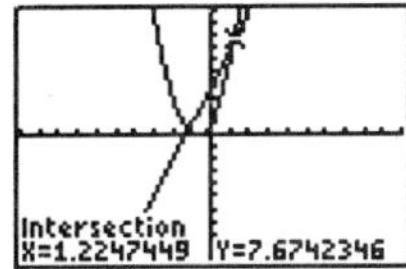

Lists – See Regression

ON/OFF

1. Press ON to turn the calculator on.

2. To turn the calculator off, press 2nd ON (OFF).

After several minutes of inactivity, the calculator will automatically shut off.

One-Variable Statistics

The 1-Variable Statistics command returns the following statistics: mean ($\bar{x}$), sum of x ($\sum x$), sum of the squares of x ($\sum x^2$), sample standard deviation (Sx), population standard deviation (σx), the amount of data entered (n), the minimum data value (minX), the first quartile (Q1), the median (Med), the third quartile (Q3), and the maximum data value (maxX).

Example: Determine all of the one-variable statistics for the following data set.

The ACT scores for some students in a particular school is given.

Student	1	2	3	4	5	6	7	8	9
ACT Score	20	22	30	32	28	21	20	34	18

1. Press STAT and select EDIT (option 1).

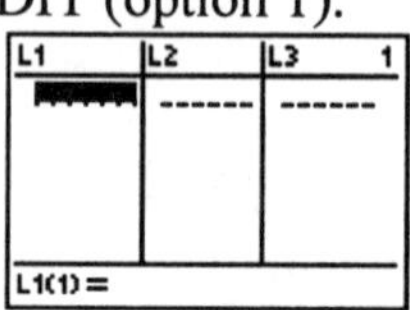

This will take you to the TI version of a spreadsheet called Lists. If data is already entered into the lists, you can clear it by using the UP ARROW to highlight the L_1 at the very top of the screen. When highlighted, press CLEAR then ENTER. Repeat for L_2, etc.

Enter the data into the lists by typing in the appropriate numbers and pressing ENTER after each entry. For this example, enter only the ACT scores in L_1.

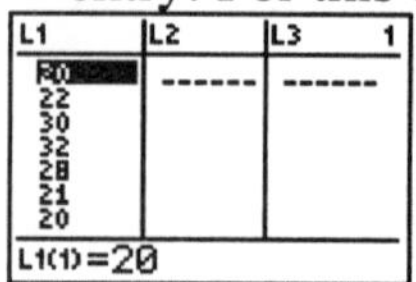

2. Press STAT; use the RIGHT ARROW to access the CALC menu. Select option 1:1-Var Stats. Press ENTER.

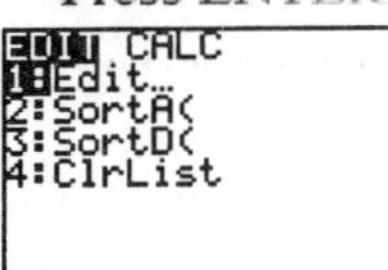

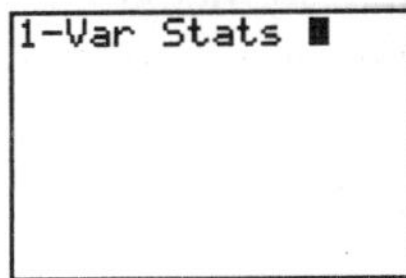

3. Notice the arrow at the bottom of the screen. This indicates that there is more information to be seen. Use the DOWN ARROW to view this information.

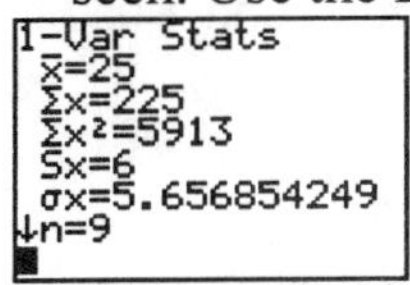

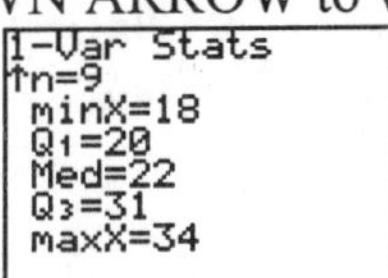

QUIT

This returns the calculator to the home screen from any other menu.

Example: Imagine that you are in the WINDOW screen and need to get back to the home screen.

1. Press WINDOW. To go back to the home screen, press 2nd MODE (QUIT).

Regression

The regression feature creates a mathematical model for data that has been entered in the Lists. For example, create a scatter plot and a mathematical model for the set of data given.

The revenue for Dell Computer Corporation from 1985 through 1993 is listed in the table. The revenue is given in millions of dollars.

Year	1985	1986	1987	1988	1989	1990	1991	1992	1993
Revenue	33.7	69.5	159.0	257.8	388.6	546.2	889.0	2013.9	2873.2

(Source: Dell Computer Corporation)

1. Push STAT and select EDIT (option 1).

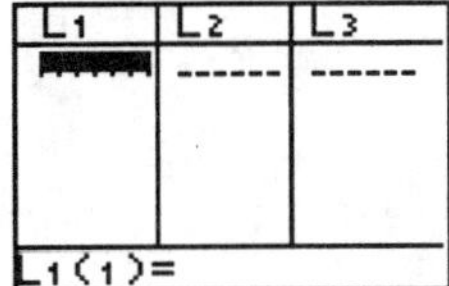

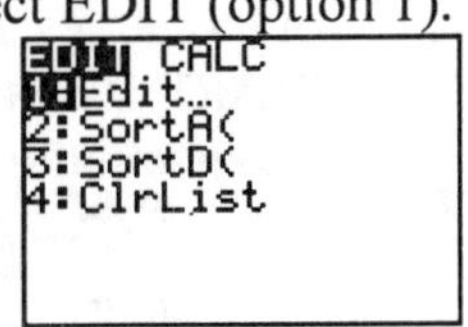

This will take you to the TI version of a spreadsheet called Lists. If data already is entered into the lists, you can clear it by using the UP ARROW to highlight the L_1 at the very top of the screen. When highlighted, press CLEAR then ENTER. Repeat for L_2, etc.

Enter the data into the lists by typing in the appropriate numbers and pressing ENTER after each entry. For this example, enter only the last two digits of the year.

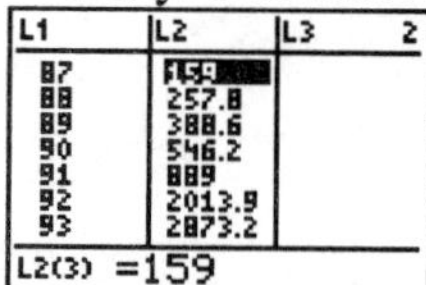

2. Set up the plot by pressing 2nd $Y =$ (STAT PLOT). Choose Plot 1, 2, or 3. Make sure the plot is turned ON, the Type is scatter plot (first option), the proper list is selected for Xlist and Ylist and the Mark is the one you want.

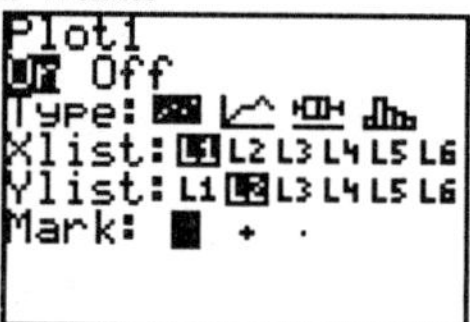

3. Adjust your WINDOW so that it reflects the values entered into your lists.

4. Press GRAPH. This graph is called a scatter plot.

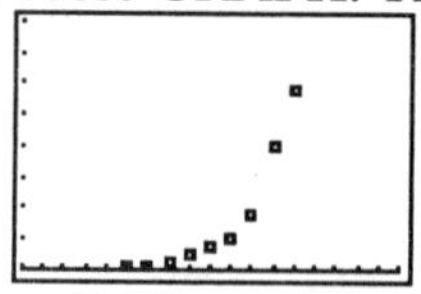

5. To have the calculator calculate a model for the data, press STAT again. This time, use your RIGHT ARROW to go over to CALC. Select the appropriate model to best fit the data.

For example: Press the DOWN ARROW to move to Option A: ExpReg. Press ENTER. This pastes the exponential regression command into the home screen. Press ENTER. The calculator returns the exponential equation in the form $Y = a*b^x$ that best fits the data.

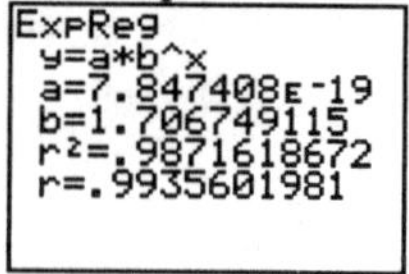

You can record this information on your paper and enter the equation in $Y =$, or…

6. To graph the function without needing to remember or write down the regression model, press $Y=$. Press VARS and select option 5: Statistics. Use the RIGHT ARROW to move across to EQ and choose option 7: RegEQ. This will paste the current regression equation into $Y =$.

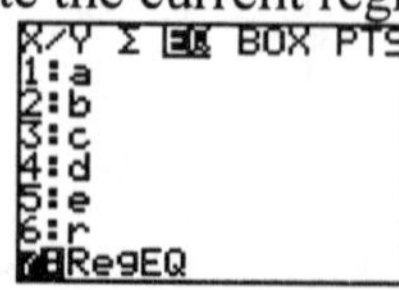

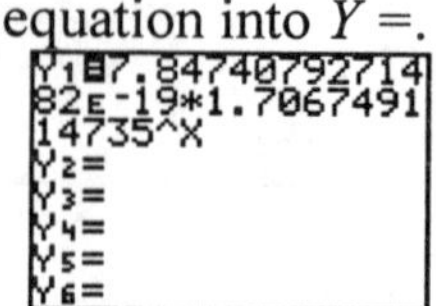

7. Press GRAPH.

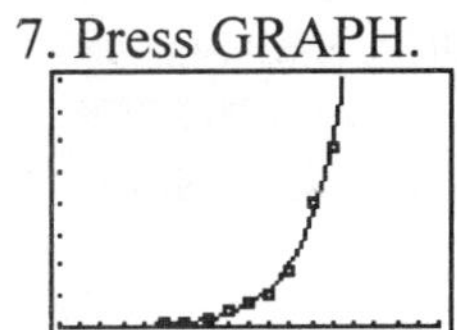

Root

The root function calculates the x-intercepts of a function in the graphing screen.

1. Enter a function in the $Y =$ menu.

   For example: $Y_1 = x^2 + 3x - 5$

2. Press ZOOM 6: ZStandard to view the function in the standard viewing window.

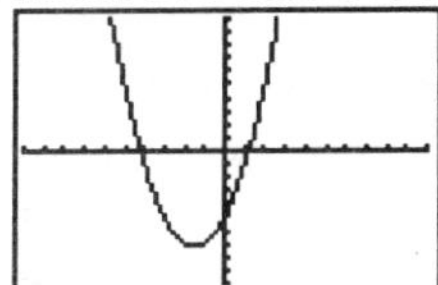

3. Press 2nd TRACE (CALC) and select option 2: root.

CALCULATE
1:value
2:root
3:minimum
4:maximum
5:intersect
6:dy/dx
7:∫f(x)dx

4. The root function calculates one x-intercept at a time. After selecting option 2: root, the calculator asks for the Lower Bound. This is any value to the left of the x-intercept being calculated. Use the LEFT and RIGHT arrow keys to move the cursor somewhere to the left of the root being calculated and press ENTER.

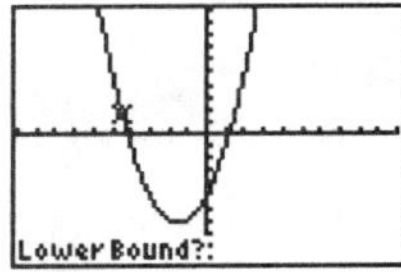

5. The calculator asks for the Right Bound. Use the LEFT and RIGHT arrow keys to move the cursor somewhere to the right of the root being calculated and press ENTER.

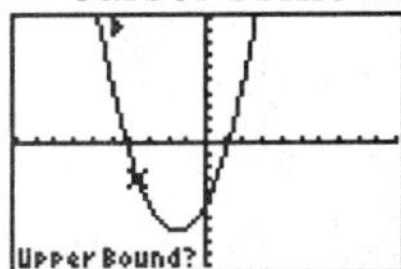

6. The calculator asks for a guess as to where the x-intercept (root) is located. Use the LEFT and RIGHT arrow keys to move the cursor as close to the x-intercept as possible and press ENTER. The x-intercept should now be displayed. Notice the y-value. It should be zero or at least very close to zero!

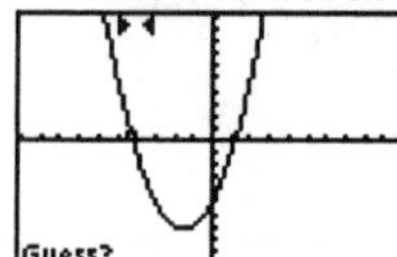

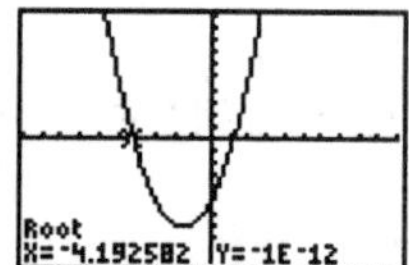

Note: In this example the y-value is -1×10^{-12}, which is extremely, close to zero!
To find the other root, repeat steps 3-6 focusing on the x-intercept on the positive x-axis.

Scatter Plots – See Regression

Scientific Notation

The calculator will automatically go in to scientific notation when the result of an operation is either very small or very large.

Example: Your old automobile hit 200,000 miles on its odometer. How many centimeters is that?

1. Enter 200,000*5280*12*2.54 then press ENTER. (Note: There are 5280 feet in one mile, 12 inches in one foot, and 2.54 centimeters in one inch.)

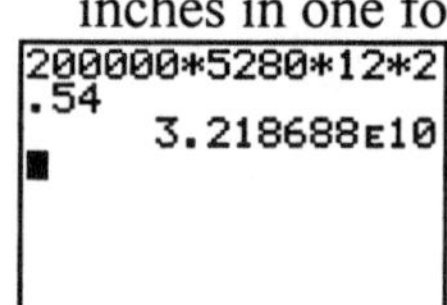

2. Notice the notation used by the calculator. The 3.210688E10 means 3.210688×10^{10}, or 32,106,880,000 centimeters.

Square Root

The calculator will calculate the square root of a number and return the positive root, accurate up to 11 decimal places.
Example: What is the square root of 111?

1. Enter the square root function first. Press 2nd x^2 ($\sqrt{\ }$). Type then the number for which you wish to calculate the square root, in this 111. Press ENTER.

```
√111
      10.53565375
```

Note: The TI-83 automatically opens the left parenthesis. Closing the parentheses is optional when taking the square root of a single number, but it is a good practice. Parentheses are needed if you need to calculate the square root of the sum or difference of numbers.

Example: What is the square root of 23 + 79?

2. Press 2nd x^2 ($\sqrt{\ }$). Enter 23+79 and close the parentheses. Press ENTER.

```
√(23+79)
         10.09950494
```

Tables

The table feature allows the user to view a table of values when the equation of the function is entered in the $Y =$ menu.

1. Enter an equation into Y_1.

   Example: $Y_1 = 2x^2 + 3x + 1$

2. Press 2nd WINDOW (TBLSET).

```
TABLE SETUP
 TblStart=5
 ΔTbl=1
Indpnt: Auto Ask
Depend: Auto Ask
```

3. TblStart determines where the table values will begin.

   Example: TblStart=0

4. ΔTbl determines the interval between table entries. For example, if increments of one are desired, enter a 1 here.

   Example: ΔTbl=1
   Note: Δ is the Greek letter delta, which stands for change. So this is literally the change in the table.

5. To view a complete table of values which can be scrolled through up or down, leave the Indpnt and Depend options on Auto.

6. To view the table, press 2nd GRAPH (TABLE).

X	Y1
0	1
1	6
2	15
3	28
4	45
5	66
6	91

X=0

7. You can scroll up or down through the table by pressing and holding the UP or DOWN arrow keys. You must have the cursor in the X column in order to use the scrolling feature.

NOTE: The values in the Y1 column of the table may be rounded. To get better accuracy, arrow over to the value in the Y1 column and look at the bottom of the screen.

Another way to evaluate a function is to change the Indpnt option to Ask.

8. Repeat steps 2-4. This time, change the Indpnt option to Ask by using the UP, DOWN, LEFT, and RIGHT ARROWS to move the cursor to the Ask option and pressing ENTER.

9. Press 2nd GRAPH (TABLE). Enter values into the X column and the calculator will display the corresponding y-value.

Example: Find the value of the function when $x = 5$.

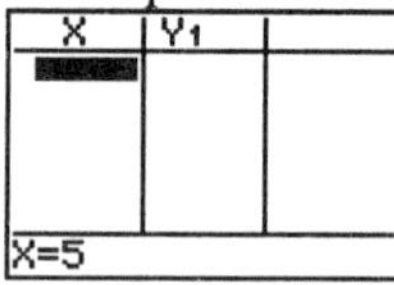

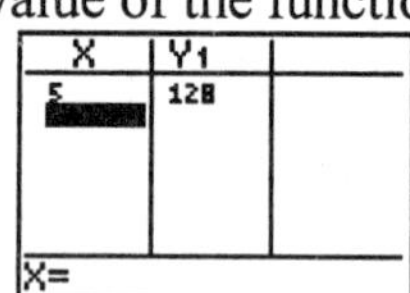

10. You can continue to evaluate the function in Y_1 by continuing to enter values for x.

TRACE

The TRACE feature allows you to trace along the graph of a function and view the coordinates of the pixels at the bottom of the screen.

Example: Graph the function $y = x^3 + 2x^2 - 3x - 4$ and use the TRACE feature to approximate the x-intercepts of the graph.

1. Enter the equation into y_1 using the $y =$ screen. Press ZOOM 6:ZStandard to graph the function in the standard viewing window.

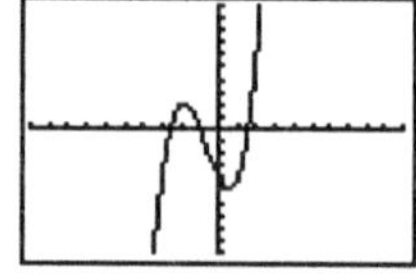

2. Press TRACE. Use the LEFT and RIGHT ARROW keys to move the cursor along the graph of the curve. Estimate the zeros of the function by moving the cursor as close to the x-intercepts as possible.

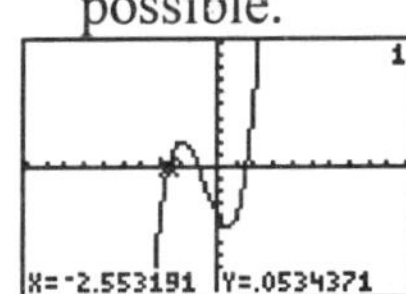

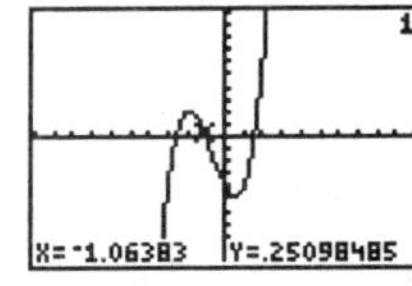

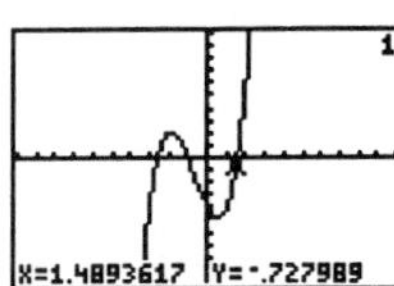

Using the TRACE feature, the zeros of this function can be estimated to be x = -2.6, -1.1, and 1.5.

Trigonometric Functions

The TI-83 is capable of calculating the sine, cosine, and tangent ratios. The inverse trigonometric functions can also be calculated. For this section, be sure that your calculator is in degree mode.

1. Press MODE. If necessary, use the DOWN ARROW to move to the Radian Degree line and highlight Degree by moving the cursor to Degree and pressing ENTER.

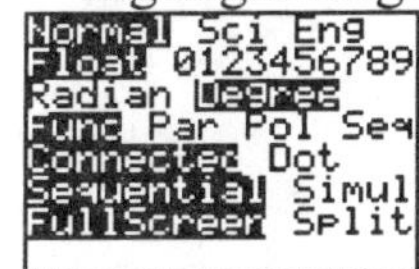

Example: Use trigonometric and inverse trigonometric functions to solve the right triangle.

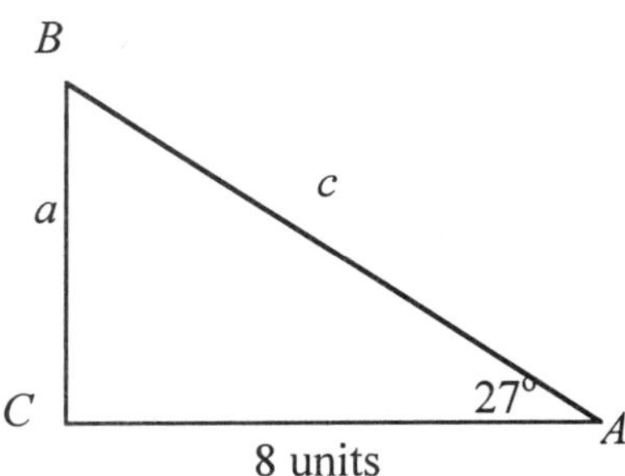

1. To determine the measure of side c, use the cosine ratio.

$$\cos 27° = \frac{8}{c}$$

$$c = \frac{8}{\cos 27°}$$

2. Press 8 ÷ COS 27 ENTER.

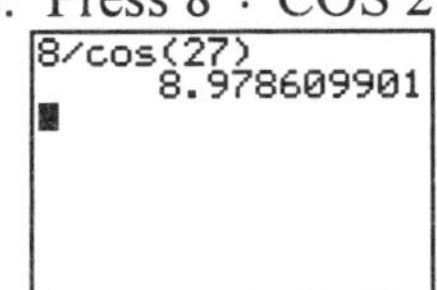

Note: The left parenthesis is not automatically inserted on the TI-82.

3. To determine the length of side a, use the tangent ratio.

$$\tan 27° = \frac{a}{8}$$

$$8 \cdot \tan 27° = a$$

4. Press 8*tan 27 ENTER.

```
8*tan(27)
          4.076203596
```

5. To determine the measure of angle B, use the inverse sine function.

$$\sin B = \frac{8}{c}$$

$$\sin B = \frac{8}{8.97}$$

$$B = \sin^{-1}\frac{8}{8.97}$$

6. Press 2nd SIN (SIN^{-1})(8 ÷ 8.97) ENTER.

NOTE: The ratio is entered in parenthesis.

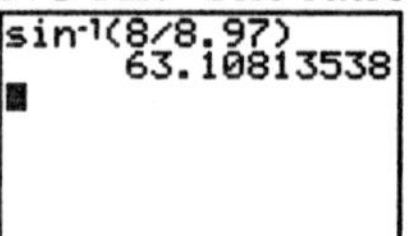

Value

The value feature can be used to evaluate a function at a particular value of x. It works in conjunction with a graph of a function.

1. Enter an equation into Y_1.

   Example: $Y_1 = 2x^2 + 3x + 1$

2. Press ZOOM 6: Zstandard to view the function in the standard viewing window.

3. To evaluate the function at different values of x, press 2nd CALC and choose option 1: value.

4. Enter a value for x for which you would like to find the value of the function.

Example: x=2

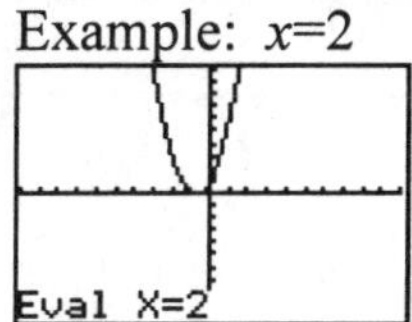

5. Press ENTER to see the value of the function at the selected value of x.

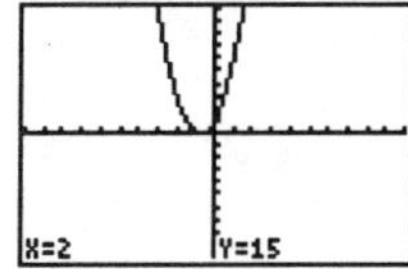

Note: You can do this many times in a row, but only for values of x in the Window Range.

Y =

The Y = menu is used to enter functions into the calculator that will be graphed.

1. Press Y =.

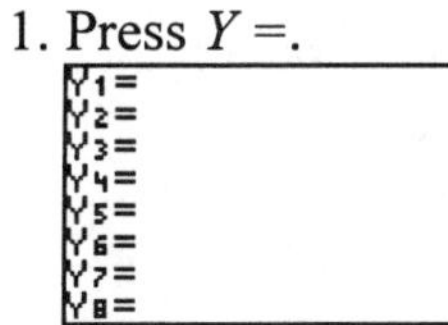

2. Up to 10 equations can be entered and graphed. Press the DOWN arrow to see Y_8-Y_0.

3. Type in the equation to be graphed.

   Example: $Y_1 = -4x + 5$ (Note: Be sure to use the grey (-) key not the blue – key when typing the $-4x$ part of the equation.)

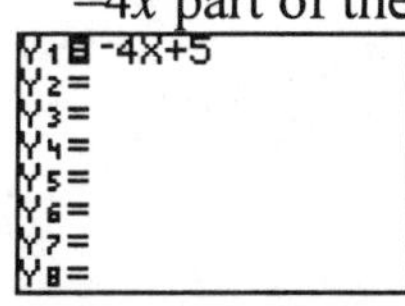

4. Press GRAPH. Depending on the viewing window, you may or may not see the graph. To view the graph in the standard viewing window, press ZOOM 6: Standard. See the ZOOM section of the manual for more details.

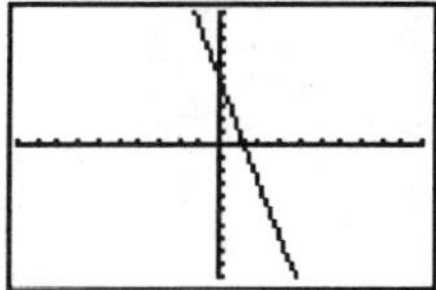

ZOOM/WINDOW

The ZOOM button changes the viewing window rapidly with pre-set ZOOM functions.

1. Enter an equation into the $Y =$ screen.

   Example: $Y_1 = 2x^2 + 3x + 1$.

2. Press ZOOM 6: ZStandard. ZStandard creates the standard viewing window which goes from –10 to 10 on the x-axis and –10 to 10 on the y-axis. Press WINDOW to see this.

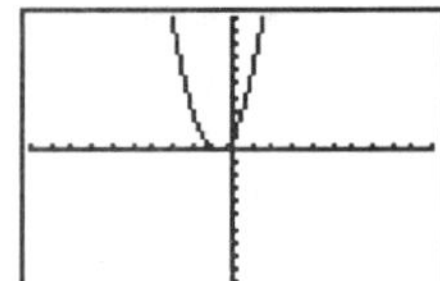
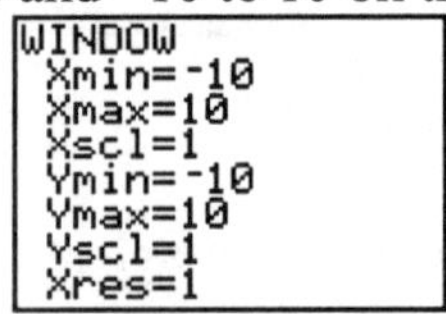

3. Press ZOOM 1: ZBox. ZBox allows zooming in on a particular region on the graph by boxing in the desired region. After pressing ZOOM 1, use the UP, DOWN, RIGHT, and/or LEFT arrow keys to move the cross hair around the screen. Think about where you would like one corner of the boxed region to be. When you have moved the cross hair to that spot, press ENTER.

   For Example:

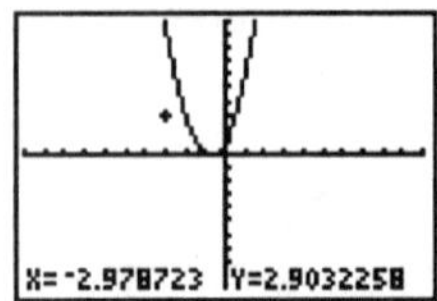

4. Use the UP, DOWN, RIGHT, and/or LEFT arrow keys to create a box around the region in which you wish to ZOOM. When the desired region is boxed in, press ENTER. When completed, press WINDOW to see how the viewing window has changed.

   For Example:

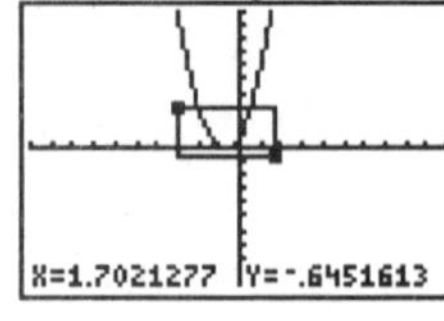

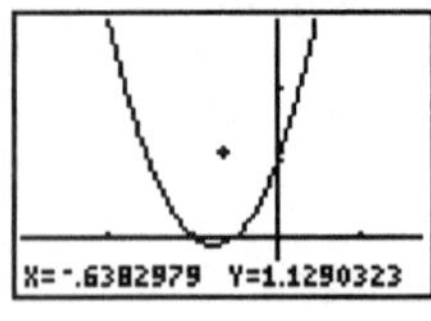

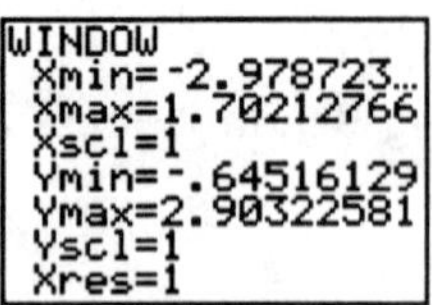

5. Press ZOOM 6: ZStandard to return the graph to the standard viewing window.

6. Press ZOOM 2: Zoom In. Zoom In will do just that, centered at a point that you select. After pressing ZOOM 2: Zoom In, use the UP, DOWN, RIGHT, and/or LEFT arrow keys to move the cross hair around the screen. Stop when the desired center of the Zoom In has been located. Press ENTER to cause the Zoom In to occur. When completed, press WINDOW to see how the viewing window has changed.

   For Example:

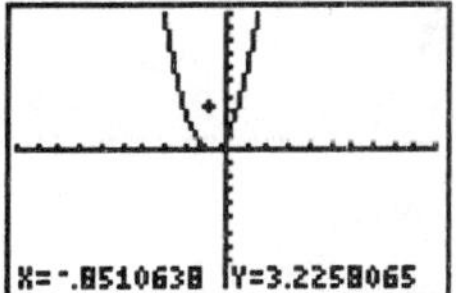

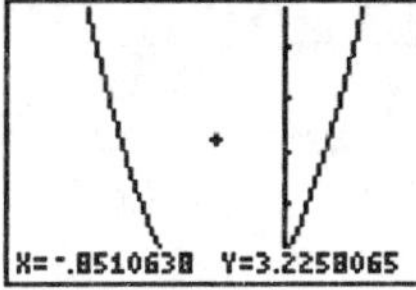

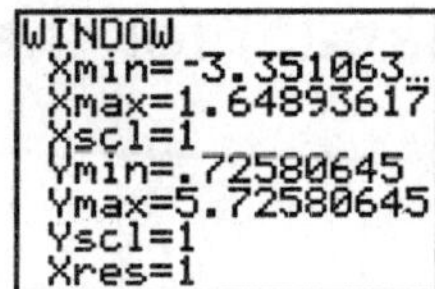

Note: The magnitude at which your calculator zooms may be different than this example.

7. Press ZOOM 6: ZStandard to return the graph to the standard viewing window.

8. Press ZOOM 3: Zoom Out. Zoom Out will do just that, centered at a point that you select. After pressing ZOOM 3: Zoom Out, use the UP, DOWN, RIGHT, and/or LEFT arrow keys to move the cross hair around the screen. Stop when the desired center of the Zoom Out has been located. Press ENTER to cause the Zoom Out to occur. When completed, press WINDOW to see how the viewing window has changed.

For Example:

X=-1.276596 Y=2.9032258

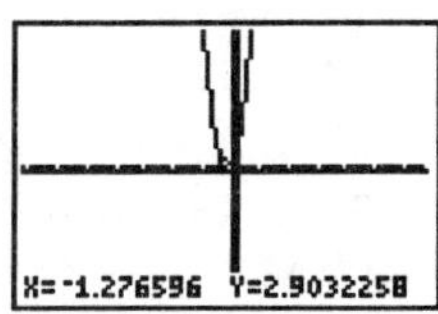

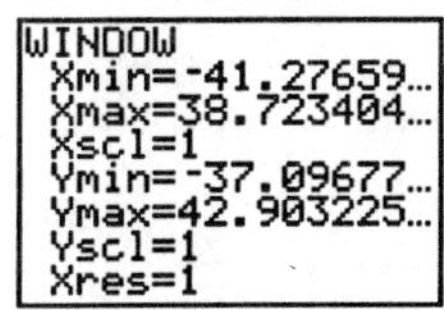

9. Press ZOOM 6: ZStandard to return the graph to the standard viewing window.

10. Press ZOOM 4: ZDecimal. ZDecimal will automatically create a viewing window that is considered “friendly.” This means that when you TRACE on a graph viewed using Zdecimal, the values shown will be “friendly” decimals, showing accuracy to the nearest tenth rather than the long decimal values shown in other viewing windows. When completed, press WINDOW to see how the viewing window has changed.

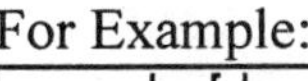
For Example:

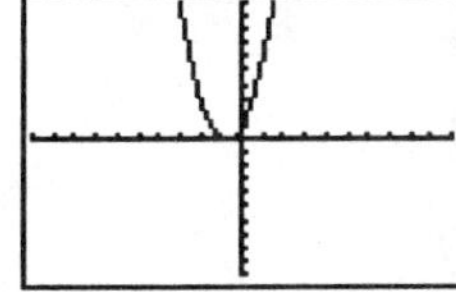

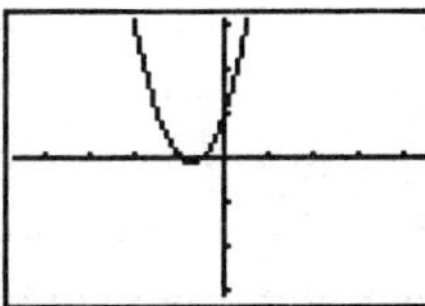
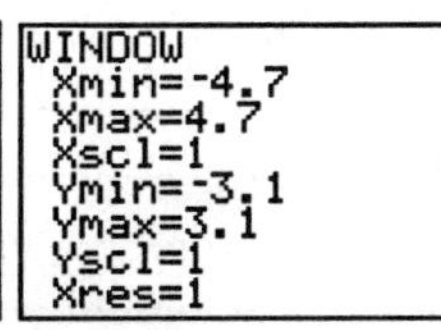

Calculator Keystroke Manual for the TI-83 and 83 Plus

Contents

Introduction

The graphing calculator is a tool to enable students to visualize mathematics, solve problems, think critically, and develop a conceptual understanding of the mathematics being studied. This Calculator Keystroke Manual is to provide students with a reference manual for using a graphing calculator. The intent is to allow students to focus on learning mathematics and for teachers to teach mathematics rather than the focus being on pushing buttons. It is written so that it can be referenced whenever a particular calculator function is desired. Each function, listed in alphabetical order, contains a step-by-step approach with calculator screen shots displaying what should be seen after each step. Occasionally, sample problems are given and worked through to better display that calculator function. Anytime a particular calculator button is referenced, the button name will be typed in all CAPS. If a 2nd function is referenced, the main button name is used with the actual function desired typed in parenthesis. For example, to turn the calculator OFF, type 2nd ON (OFF).

This manual does not provide instructions on using every function available on the calculator. With this background and subsequent mathematics courses, students will continue to develop their ability to use the calculator as an effective and efficient tool for doing mathematics.

Before beginning, be sure that the calculator is prepared correctly.

Preparing the Calculator

1. Press MODE. For the purposes of this manual, all settings on the left hand side of the screen should be highlighted. If something is not set properly, use the UP, DOWN, LEFT and RIGHT keys to move the cursor to the desired location. Press ENTER to highlight the appropriate setting.

2. Press 2nd $Y=$ (STAT PLOT). Use the DOWN arrow key to move the cursor to option 4:PlotsOff. Press ENTER.

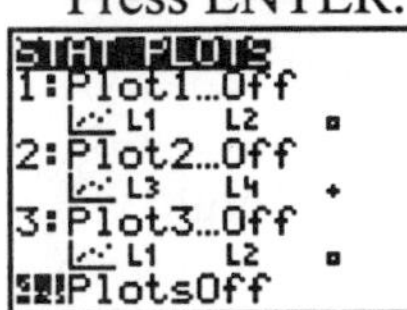

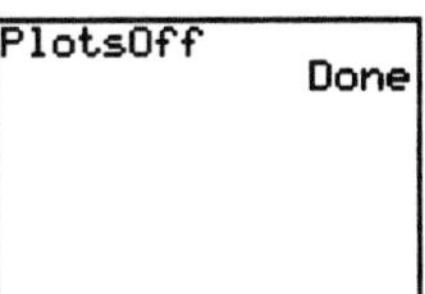

3. The HOME SCREEN is the main screen where most calculations are done.

Using This Manual

A good way to use this manual is to complete each of sections in the order given, then use the manual as a reference when needed.

1. ON/OFF
2. Contrast
3. ANS (Answer)
4. DEL/INS (Delete/Insert)
5. ENTRY
6. Square Root
7. Exponents
8. Scientific Notation
9. QUIT
10. Fractions
11. *Y* =
12. ZOOM/WINDOW
13. TRACE
14. Tables
15. Absolute Value
16. Value
17. Evaluate a Function
18. Intersect
19. ZERO
20. One-Variable Statistics
21. Histograms
22. Box and Whisker Plots
23. Regression/Scatter Plots/Lists
24. Error Messages
25. Trigonometric Functions
26. Time Value of Money (TVM)

Absolute Value

The absolute value of a number always returns the positive value of that number. The absolute value function can be performed numerically in the home screen or can be graphed as a function in the $y =$ screen.

Example: Determine the absolute value of –5.

1. From the home screen, press MATH, press the RIGHT ARROW to highlight the NUM menu, and select 1: abs(

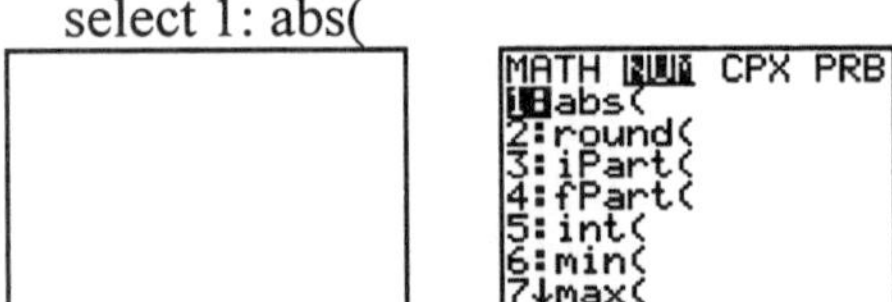

2. Enter –5 in the parentheses. Be sure to use the negative button (-) not the blue minus button.

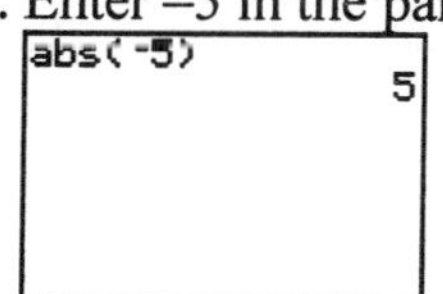

Example: Graph $Y_1 = |x|$.

3. Press $Y=$ and enter $Y_1 = \text{abs}(x)$. Follow the previous instructions for accessing the abs command.

Plot1 Plot2 Plot3
\Y1 ■abs(X)
\Y2=
\Y3=
\Y4=
\Y5=
\Y6=
\Y7=

4. Press ZOOM 6: ZStandard to graph the function in the standard viewing window.

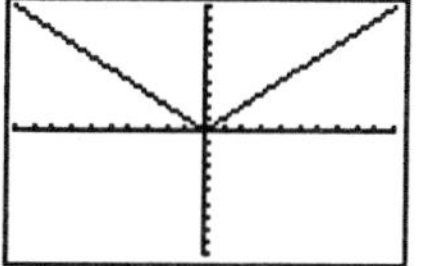

ANS

The ANS function recalls the previous answer so that additional operations can be performed on the answer to a previous operation.

Example: Start with the number 1 and continue to double it indefinitely.

1. Enter the number 1 and press ENTER. Press the multiplication button (×) then press 2 and ENTER. The ANS part of the display refers to the previous answer, in this case the 1 that was entered initially.

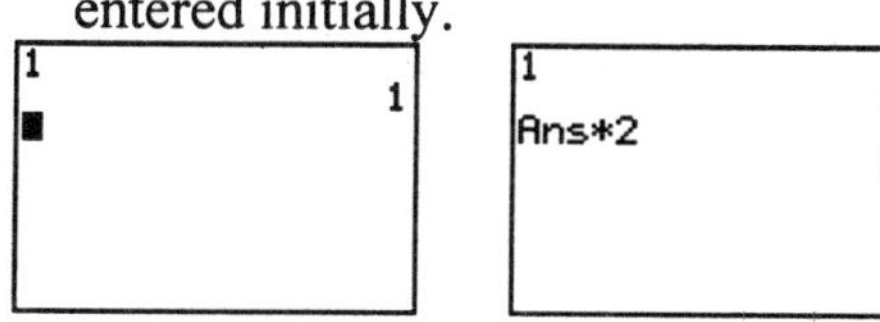

2. Continue to press ENTER until reaching the answer 64. Each time, the calculator takes the previous answer and multiples by 2.

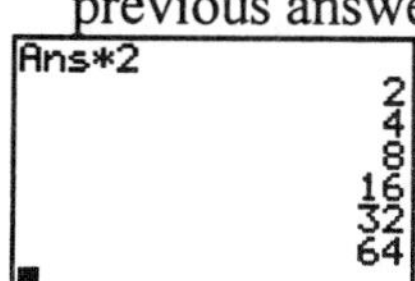

The ANS function can be accessed by typing 2nd (-) (ANS).

3. Press 7, press +, press 2nd (-) (ANS), then ENTER. The calculator takes the last answer (64), adds 7, and returns the answer of 71.

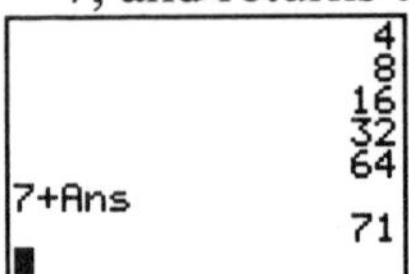

Box and Whisker Plots

Box and whisker plots are used to display and compare data. Create a box and whisker plot for the data given.

The ACT scores for some students in a particular school is given.

Student	1	2	3	4	5	6	7	8	9
ACT Score	20	22	30	32	28	21	20	34	18

1. Press STAT and select 1: EDIT.

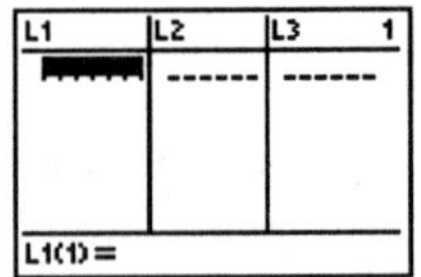

This will take you to the TI version of a spreadsheet called Lists. If data already is entered into the lists, you can clear it by using the UP ARROW to highlight the L1 at the very top of the screen. When highlighted, press CLEAR and ENTER. Repeat for L2, etc.

Enter the data into the lists by typing in the appropriate numbers and pressing ENTER after each entry. For this example, enter only the ACT scores into L1.

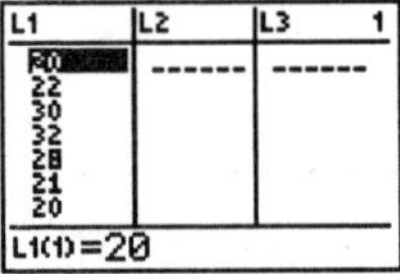

2. Set up the plot by pressing 2nd $Y =$ (STAT PLOT). Choose Plot 1, 2, or 3. Make sure that the plot is turned ON, the type is Box and Whisker (2nd to the last choice), the Xlist is the list containing the data (L_1 in this case), and the frequency (Freq) is 1.

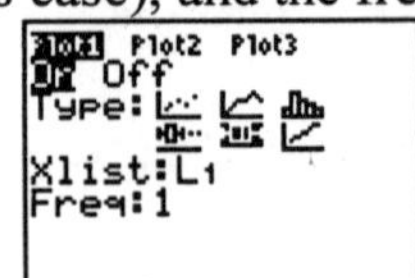

3. Adjust your WINDOW so that it reflects the values entered into the lists. For a Box and Whisker plot, only the Xmin and Xmax must reflect the data. The Ymin and Ymax are irrelevant.

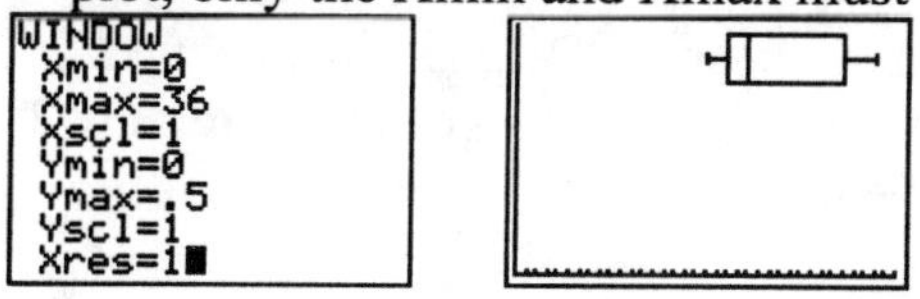

```
WINDOW
 Xmin=0
 Xmax=36
 Xscl=1
 Ymin=0
 Ymax=.5
 Yscl=1
 Xres=1
```

4. Use the TRACE command and your RIGHT and LEFT ARROW keys to find the minimum data value, first quartile, median, third quartile, and maximum data value.

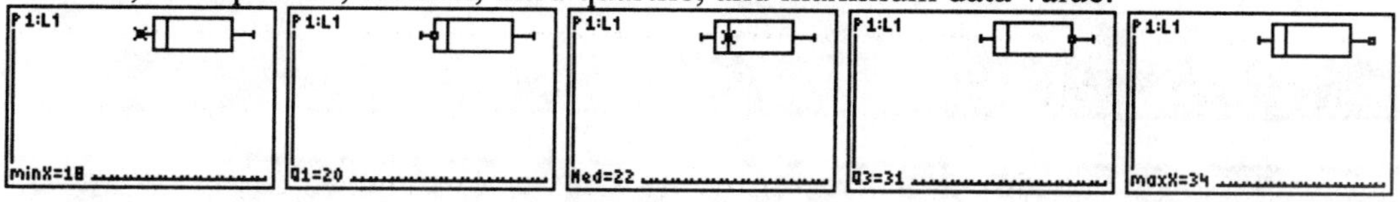

```
P1:L1  minX=18
P1:L1  Q1=20
P1:L1  Med=22
P1:L1  Q3=31
P1:L1  maxX=34
```

Contrast

The contrast for the calculator screen can be adjusted to be darker or lighter. As your batteries become older, it will become necessary to adjust the contrast so that you can more easily read the calculator screen.

1. Press 2nd and release.

2. Hold the DOWN ARROW key to make the screen lighter. Hold the UP ARROW key to make the screen darker. Each time the DOWN or UP ARROW key is released, the 2nd key must be pressed again to continue to adjust the contrast.

3. As the UP or DOWN ARROW keys are held, notice the number in the upper right hand corner. The contrast is measured from 1 – 9 where 1 is the lightest setting and 9 is the darkest. If the contrast needs to be 9 in order for you to see the screen, you probably need to replace the batteries.

DEL/INS

The DEL (Delete) key deletes the character under the current cursor location. The 2nd DEL (INS) (Insert) key inserts a character to the left of the current cursor location.

Example: You desire to type 9+6-5*34, but accidentally type 9+*36*-5**4*.

1. Enter 9+*36*-5**4*. Use the LEFT arrow to move the cursor to the 3 location.

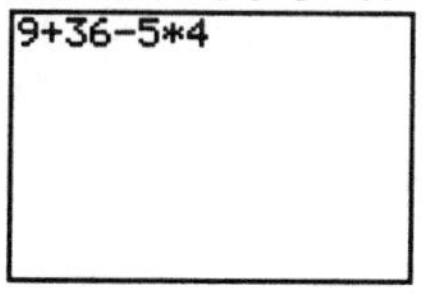

```
9+36-5*4
```

2. Press DEL. This deletes the unneeded 3.

3. Use the RIGHT ARROW to move the cursor to the 4 location. Press 2nd DEL (INS) and enter a 3. This inserts the 3 to the left of the 4 as desired. Notice that the cursor changed after pressing 2nd DEL (INS) so that you knew that the calculator was in the insert mode.

9+6-5*_

9+6-5*34■

ENTRY

The ENTRY function returns previously entered operations on the calculator screen so that you may edit them.

1. Press 2nd ENTER (ENTRY). The last operation entered into the computer appears on the screen and the cursor is at the right end of the line. Use the UP, DOWN, RIGHT, LEFT arrow keys to edit the line.

2. Press 2nd ENTER (ENTRY) again. Continue to press 2nd ENTER (ENTRY) and notice that previous entries continue to be restored on the screen. Any one of them can be edited.

Error Messages

Occasionally, error messages are given to indicate some problem with the function you are asking the calculator to perform. Three common error messages and how to correct them are provided.

Syntax Error
Syntax refers to the way in which the function or command was entered. One common error deals with use of parentheses.

Example: You intend to enter 6*(3+2) but accidentally enter 6*3+2).

1. In the Home Screen, enter 6*3+2) and press ENTER.

6*3+2)■

ERR:SYNTAX
1:Quit
2:Goto

2. Because of the missing opening parenthesis, the calculator returns a syntax error. Notice the options on the screen. You can QUIT and start over, or you can GOTO the error so that it can be corrected. Select GOTO by pressing ENTER or the number 2.

3. To correct the mistake, insert a parenthesis before the 3 using INS. Press ENTER.

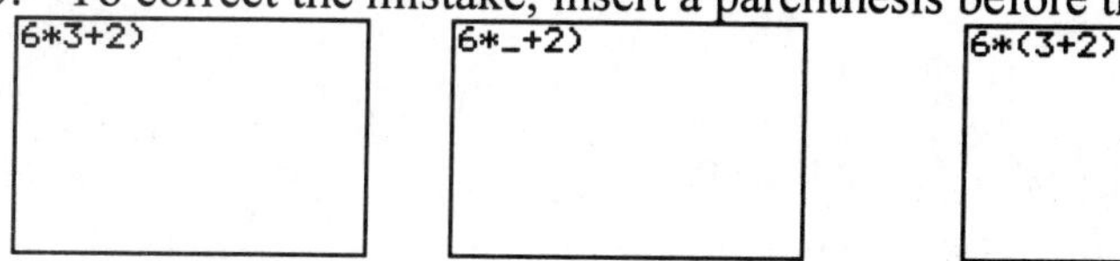

Window Range Error

The Graph Window Error refers to an incorrect WINDOW entry.

1. Enter an equation into $Y =$.

   For example: $Y_1 = x^3 + 3$

2. Press ZOOM 6: ZStandard to graph the function in the standard viewing window.

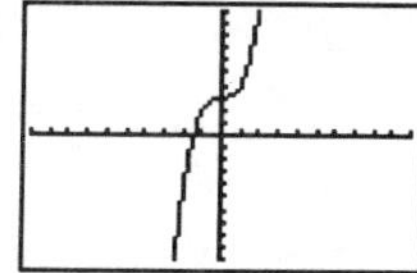

3. Press WINDOW. Imagine that you incorrectly changed the WINDOW as shown below. Press GRAPH. A WINDOW RANGE Error occurs.

4. The only option is to QUIT. To correct the error, press WINDOW and change the Xmin and Xmax as shown. Press GRAPH.

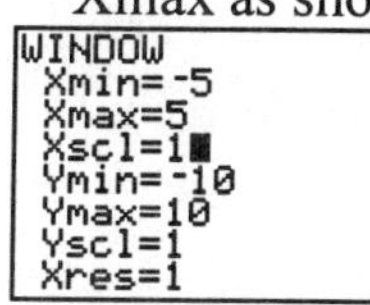

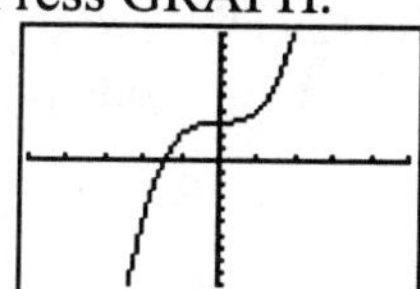

Dimension Mismatch Error (DIM MISMATCH)

The Dimension Mismatch error results when the data entered in the Lists do not match up. That is, there is more data entered in one list than in the other.

1. When DIM MISMATCH is encountered, the only option is to QUIT (Option 1).

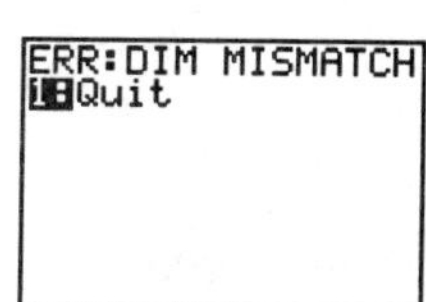

2. Press QUIT. To correct the error, refer to Preparing the Calculator, step 2.

Evaluate a Function

One way to evaluate a function is to use the notation $Y_1(x)$ in the home screen.

1. Enter an equation into Y_1.

   Example: $Y_1 = 2x^2 + 3x + 1$

2. Press 2nd MODE (QUIT) to return to the home screen.

3. Press VARS, arrow across to the right to Y-VARS and select option 1:Function.

4. Since the equation was entered in Y_1, select option 1: Y_1. Type a left parenthesis, enter a value for x, close the parenthesis and press ENTER.

   Example: $Y_1(2)$

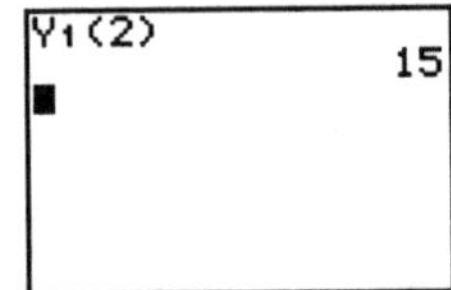

5. If many evaluations need to be completed, press 2nd ENTER (ENTRY) to repeat the previous entry, then use arrows to move the cursor over the value of x that needs to be changed and type in a new value for x and press ENTER.

Exponents

Exponents of power 2 can be entered into the calculator 2 different ways. Exponents of 2 or greater are entered using the caret key.

Example: Evaluate $3^2 + 7^2 - 8^4 + 2^7$.

1. To enter the powers of 2, use the x^2 key that is located in the middle of the first column of keys. To enter powers of 2 or greater, use the caret key (∧) which is located below the CLEAR key in the last column.

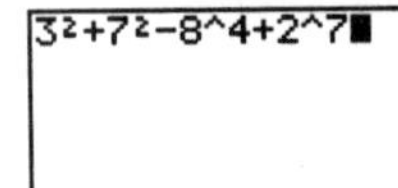

2. Press ENTER.

```
3²+7²-8^4+2^7
           -3910
■
```

Fractions

The TI-83 can convert from fractions to decimals and from decimals to fractions.

Example: Add $\frac{3}{8}+\frac{5}{12}$. Write your answer as a fraction.

1. Enter 3/8 + 5/12 and press ENTER. The fraction bar (/) is the division symbol on the keypad.

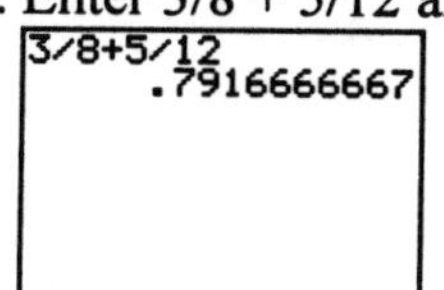

2. To change the answer from a decimal to a fraction, press MATH and select option 1: ▶ Frac.

```
3/8+5/12
     .7916666667
Ans▶Frac
           19/24
```

Note: See the ANS section if you don't know what Ans means.

3. To convert the fraction back to a decimal, press MATH and select option 2: ▶ Dec.

```
3/8+5/12
     .7916666667
Ans▶Frac
           19/24
Ans▶Dec
     .7916666667
■
```

This screen shows that the fraction $\frac{19}{24}$ is equivalent to the decimal 0.7916666667.

Histograms

Histograms are a way to display and analyze data. Create a histogram for the data given.

The shoe size of some the children in the class is shown.

Child	1	2	3	4	5	6	7	8	9
Shoe Size	5	5	4	5	4	3	6	5	4

1. Press STAT and select 1: EDIT.

This will take you to the TI version of a spreadsheet called Lists. If data already is entered into the lists, you can clear it by using the UP ARROW to highlight the L1 at the very top of the screen. When highlighted, press CLEAR and ENTER. Repeat for L2, etc.

Enter the data into the lists by typing in the appropriate numbers and pressing ENTER after each entry. For this example, enter only the shoe sizes into L1.

2. Set up the plot by pressing 2nd $Y =$ (STAT PLOT). Choose Plot 1, 2, or 3. Make sure that the plot is turned ON, the type is Histogram (3rd choice), the Xlist is the list containing the data (L_1 in this case), and the frequency (Freq) is 1.

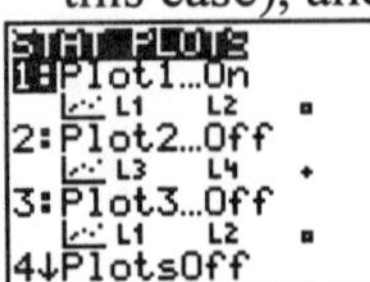

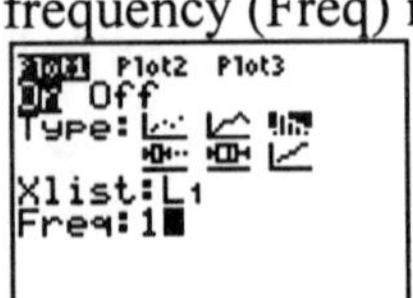

3. Adjust your WINDOW so that it reflects the values entered into the lists. The Ymin should be set at 0 because 0 is the least number of times a particular shoe size is seen. Ymax should be set at 5 since the largest number of times a particular shoe size is seen is 4 (4 children wear size 5 shoes).

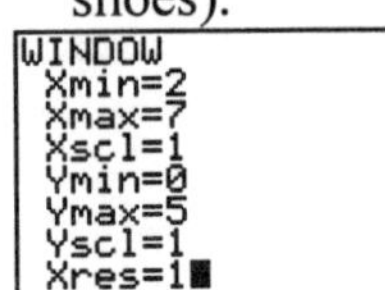

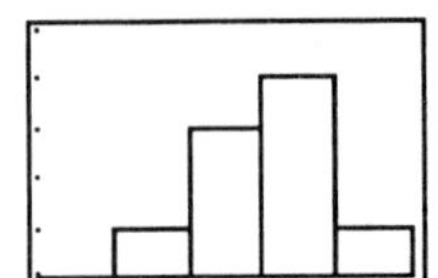

4. Use the TRACE command and your RIGHT and LEFT ARROW keys to find the frequency of each shoe size.

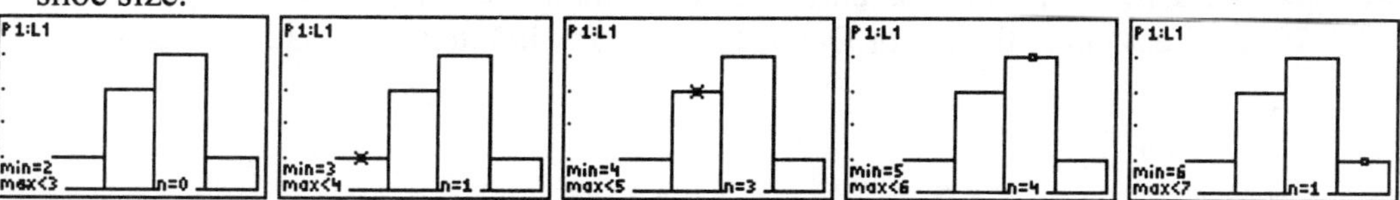

Intersect

The Intersect function will calculate an intersection point of two graphed equations.

1. Enter the equations in Y_1 and Y_2.

Example: $Y_1 = 2x^2 + 3x + 1$
$Y_2 = 3x + 4$

2. Select an appropriate viewing window. In this case the standard viewing window may be used (ZOOM 6).

3. Either press GRAPH or ZOOM 6.

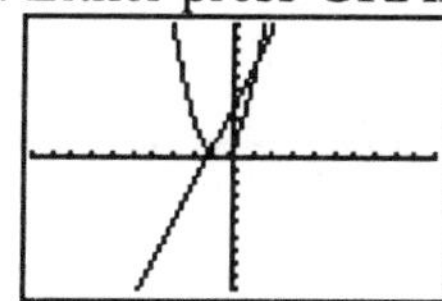

4. To determine the first quadrant point of intersection for these two graphs, press 2nd TRACE (CALC) and select option 5 (intersect).

5. Automatically returning to the graph screen, the calculator asks for the First curve. The First curve is the function defined in Y_1. At the top of the screen, the equation entered into Y_1 should be showing. If it is, press ENTER. If not, press either the UP or DOWN arrow to switch to that equation, then press ENTER.

6. The calculator now asks for the Second curve. The Second curve is the function defined in Y_2. At the top of the screen, the equation entered into Y_2 should be showing. If it is, press ENTER. If not, press either the UP or DOWN arrow to switch to that equation, then press ENTER.

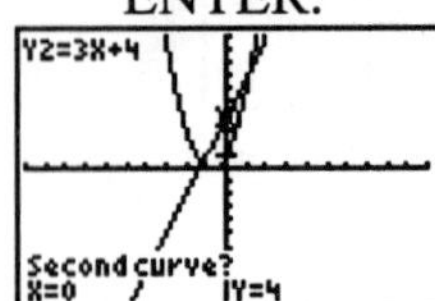

7. The calculator next asks for a guess as to where the point of intersection you are looking for is located on the graph. Use the LEFT and RIGHT arrow keys to move the cursor to the place where the graphs intersect in the first quadrant (in this example).

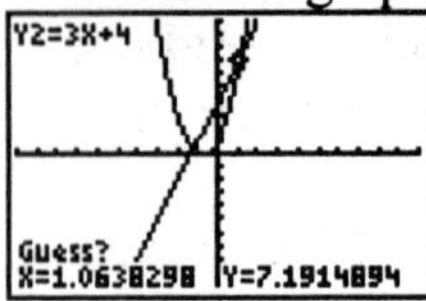

8. The calculator will show the point of intersection at the bottom of the screen.

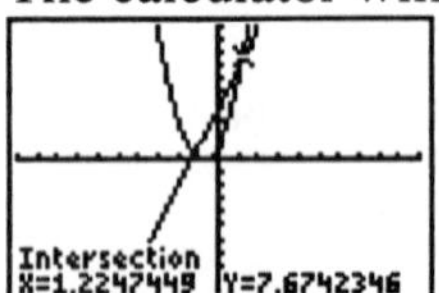

Lists – See Regression

ON/OFF

1. Press ON to turn the calculator on.

2. To turn the calculator off, press 2nd ON (OFF).

After several minutes of inactivity, the calculator will automatically shut off.

One-Variable Statistics

The 1-Variable Statistics command returns the following statistics: mean ($\bar{x}$), sum of x ($\sum x$), sum of the squares of x ($\sum x^2$), sample standard deviation (Sx), population standard deviation (σx), the amount of data entered (n), the minimum data value (minX), the first quartile (Q1), the median (Med), the third quartile (Q3), and the maximum data value (maxX).

Example: Determine all of the one-variable statistics for the following data set.

The ACT scores for some students in a particular school is given.

Student	1	2	3	4	5	6	7	8	9
ACT Score	20	22	30	32	28	21	20	34	18

1. Press STAT and select EDIT (option 1).

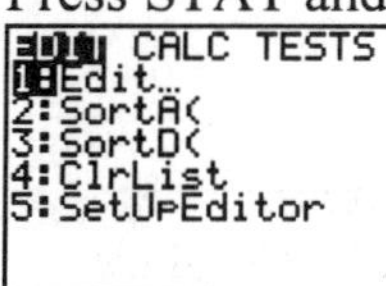

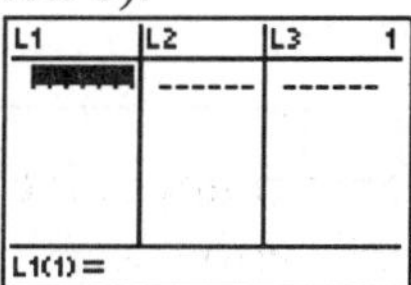

This will take you to the TI version of a spreadsheet called Lists. If data is already entered into the lists, you can clear it by using the UP ARROW to highlight the L_1 at the very top of the screen. When highlighted, press CLEAR then ENTER. Repeat for L_2, etc.

Enter the data into the lists by typing in the appropriate numbers and pressing ENTER after each entry. For this example, enter only the ACT scores in L_1.

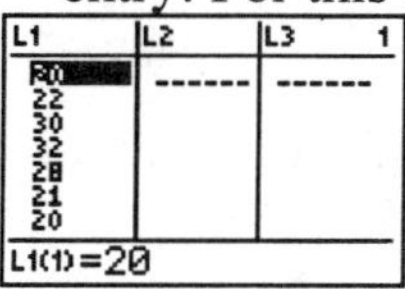

2. Press STAT; use the RIGHT ARROW to access the CALC menu. Select option 1:1-Var Stats. Press ENTER.

3. Notice the arrow at the bottom of the screen. This indicates that there is more information to be seen. Use the DOWN ARROW to view this information.

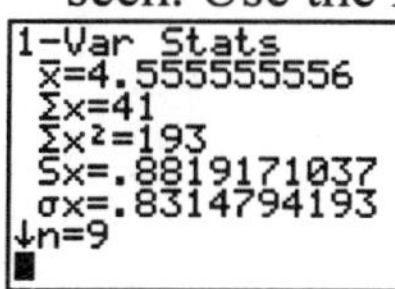

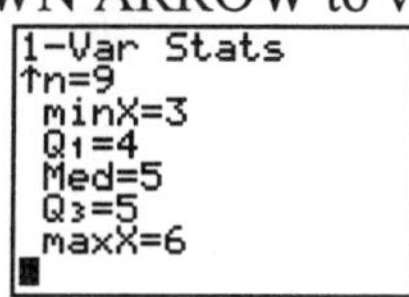

QUIT

This returns the calculator to the home screen from any other menu.

Example: Imagine that you are in the WINDOW screen and need to get back to the home screen.

1. Press WINDOW. To go back to the home screen, press 2nd MODE (QUIT).

Regression

The regression feature creates a mathematical model for data that has been entered in the Lists. For example, create a scatter plot and a mathematical model for the set of data given.

The revenue for Dell Computer Corporation from 1985 through 1993 is listed in the table. The revenue is given in millions of dollars.

Year	1985	1986	1987	1988	1989	1990	1991	1992	1993
Revenue	33.7	69.5	159.0	257.8	388.6	546.2	889.0	2013.9	2873.2

(Source: Dell Computer Corporation)

1. Press STAT and select EDIT (option 1).

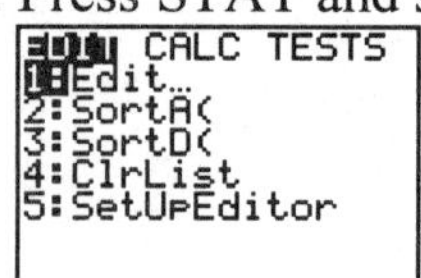

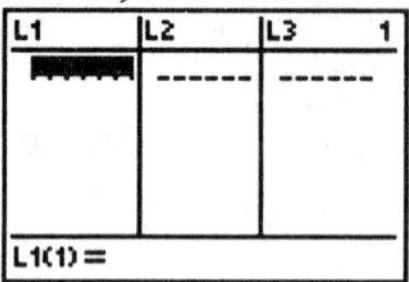

This will take you to the TI version of a spreadsheet called Lists. If data is already entered into the lists, you can clear it by using the UP ARROW to highlight the L_1 at the very top of the screen. When highlighted, press CLEAR then ENTER. Repeat for L_2, etc.

Enter the data into the lists by typing in the appropriate numbers and pressing ENTER after each entry. For this example, enter only the last two digits of the year.

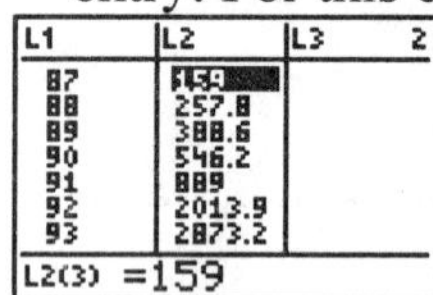

2. Set up the plot by pressing $2^{nd}Y$ = (STAT PLOT). Choose Plot 1, 2, or 3. Make sure the plot is turned ON, the type is scatter plot, the proper list is selected for Xlist and Ylist and the Mark is the one you want. To name your Xlist L_1 (if it isn't already), type 2nd and the number 1 button.

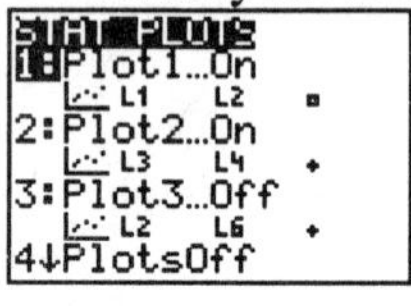

3. Adjust your WINDOW so that it reflects the values entered into your lists.

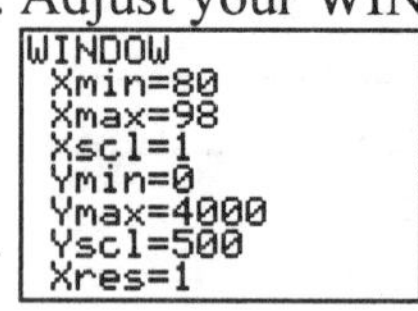

4. Press GRAPH. This graph is called a scatter plot.

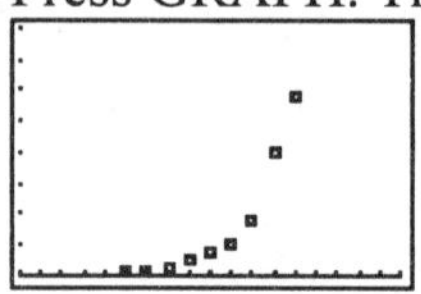

5. To have the calculator calculate a model for your data, press STAT again. This time, use your RIGHT ARROW to go over to CALC. Select the appropriate model to best fit the data.

For example: Press the DOWN ARROW to move down to Option 0: ExpReg. Press ENTER. This pastes the exponential regression command into the home screen. Press ENTER. The calculator returns the exponential equation in the form $y = a{\cdot}b^x$ that best fits the data.

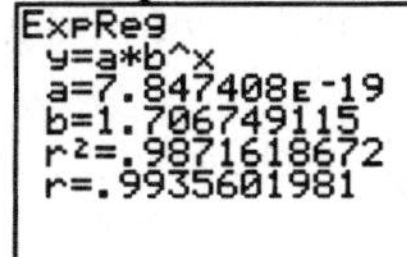

You can record this information on your paper and enter the equation in y =, or…

6. To graph the function without needing to remember or write down the regression model, press Y=. Press VARS and select option 5: Statistics. Use the RIGHT ARROW to move across to EQ and choose option 1: RegEQ. This will paste the current regression equation into Y =.

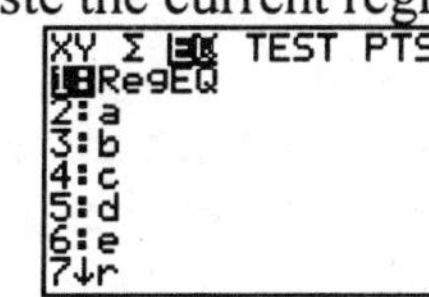
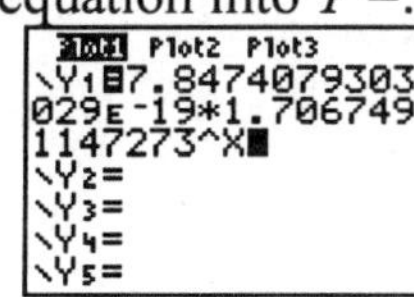

7. Press GRAPH.

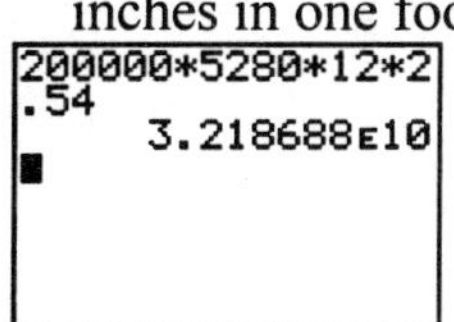

Scatter Plots – See Regression

Scientific Notation

The calculator will automatically go in to scientific notation when the result of an operation is either very small or very large.

Example: Your old automobile hit 200,000 miles on its odometer. How many centimeters is that?

1. Enter 200,000*5280*12*2.54 then press ENTER. (Note: There are 5280 feet in one mile, 12 inches in one foot, and 2.54 centimeters in one inch.)

```
200000*5280*12*2
.54
      3.218688E10
```

2. Notice the notation used by the calculator. The 3.210688E10 means 3.210688 x 10^{10}, or 32,106,880,000 centimeters.

Square Root

The calculator will calculate the square root of a number and return the positive root, accurate up to 11 decimal places.

Example: What is the square root of 111?

1. Enter the square root function first. Press 2nd x^2 ($\sqrt{\ }$). Type the number for which you wish to calculate the square root, in this 111. Press ENTER.

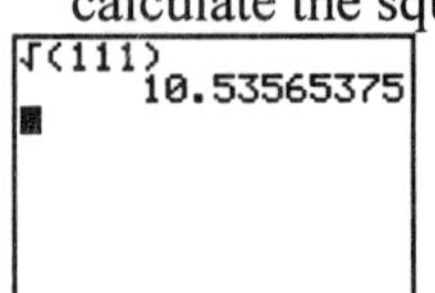

Note: The TI-83 automatically opens the left parenthesis. Closing the parentheses is optional when taking the square root of a single number, but it is a good practice. Parentheses are needed if you need to calculate the square root of the sum or difference of numbers.

Example: What is the square root of 23 + 79?

2. Press 2nd x^2 ($\sqrt{\ }$). Enter 23+79 and close the parentheses. Press ENTER.

```
√(23+79)
          10.09950494
```

Tables

The table feature allows the user to view a table of values when the equation of the function is entered in the $Y=$ menu.

1. Enter an equation into Y_1.

   Example: $Y_1 = 2x^2 + 3x + 1$

2. Press 2nd WINDOW (TBLSET).

3. TblStart determines where the table values will begin.

   Example: TblStart=0

4. ΔTbl determines the interval between table entries. For example, if increments of one are desired, enter a 1 here.

   Example: ΔTbl=1

   Note: Δ is the Greek letter delta, which stands for change. So this is literally the change in the table.

5. To view a complete table of values which can be scrolled through up or down, leave the Indpnt and Depend options on Auto.

6. To view the table, press 2nd GRAPH (TABLE).

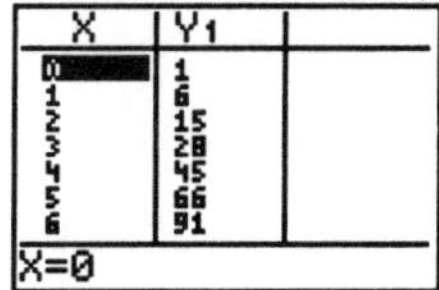

7. You can scroll up or down through the table by pressing and holding the UP or DOWN arrow keys. You must have the cursor in the X column in order to use the scrolling feature.

   NOTE: The values in the Y1 column of the table may be rounded. To get better accuracy, arrow over to the value in the Y1 column and look at the bottom of the screen.

Another way to evaluate a function is to change the Indpnt option to Ask.

8. Repeat steps 2-4. This time, change the Indpnt option to Ask by using the UP, DOWN, LEFT, and RIGHT ARROWS to move the cursor to the Ask option and pressing ENTER.

9. Press 2nd GRAPH (TABLE). Enter values into the X column and the calculator will display the corresponding y-value.

   Example: Find the value of the function when $x = 5$.

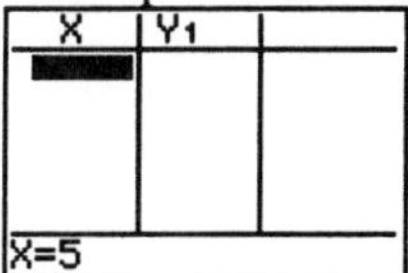

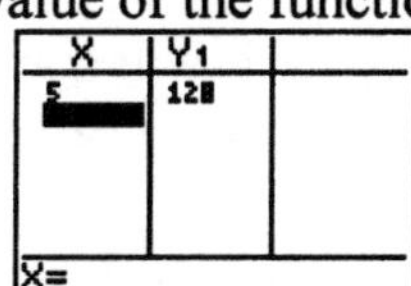

10. You can continue to evaluate the function in Y_1 by continuing to enter values for x.

Time Value of Money - TVM Solver

The TVM (Time Value of Money) Solver can be used to calculate a house or car payment, the interest rate of a loan, or other financial options. The TVM is not available on the TI-82, 85, or 86.

For example, consider the following situation. A house is purchased for $85,000 financed for 30 years at 8.5% interest. What will the monthly payment be on this mortgage?

1. Press 2nd x^{-1} (FINANCE). Select Option 1:TVM Solver….

NOTE FOR TI-83 PLUS CALCULATORS: To access the FINANCE menu, press APPS and select Option 1: Finance. Then select Options 1:TVM Solver.

2. Enter *N* (number of payments). In this situation, *N* = 360 since the mortgage is for 30 years and 12 payments will be made each year.

3. Enter *I%*. This is the interest rate and should be entered as a percentage. In this situation, *I%*=8.5.

4. Enter *PV*. This is the present value of the loan, 85000 in this situation.

5. Skip *PMT* and arrow down to *FV*. Enter *FV*. This is the future value of the loan which is 0 (the loan is paid off).

6. Enter *P/Y*. This is the number of payments per year. In this situation, payments will be made each month, so *P/Y* =12.

7. Enter *C/Y*. This is the number of compounding periods per year. Use *C/Y* =12 for this situation.

8. Select *PMT*. This tells when the payment will be made, at the end (END) or the beginning (BEGIN) of the month. For this situation, use BEGIN.

9. To calculate the monthly payment in this situation, arrow up to the *PMT*: line. Press ALPHA ENTER (SOLVE) and the monthly payment will be calculated. It will appear as a negative value because the calculator considers it a payment made (outflow of cash).

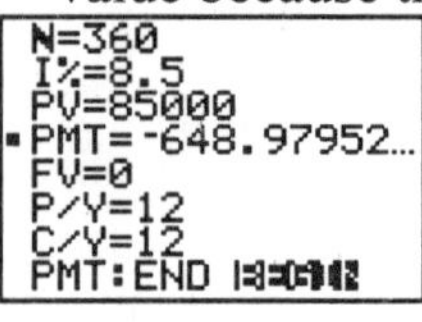

10. If other information was unknown rather than the monthly payment, simply enter the known values, move the cursor to the unknown value and press ALPHA ENTER (SOLVE) to solve for that unknown value.

TRACE

The TRACE features allows you to trace along the graph of a function and view the coordinates of the pixels at the bottom of the screen.

Example: Graph the function $y = x^3 + 2x^2 - 3x - 4$ and use the TRACE feature to approximate the x-intercepts of the graph.

1. Enter the equation into Y_1 using the $Y =$ screen. Press ZOOM 6:ZStandard to graph the function in the standard viewing window.

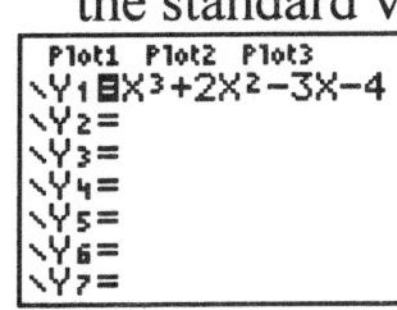

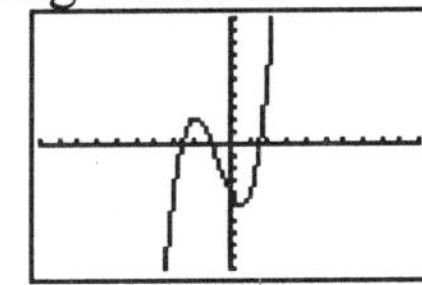

2. Press TRACE. Use the LEFT and RIGHT ARROW keys to move the cursor along the graph of the curve. Estimate the zeros of the function by moving the cursor as close to the x-intercepts as possible.

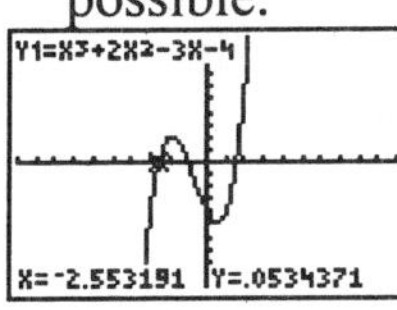

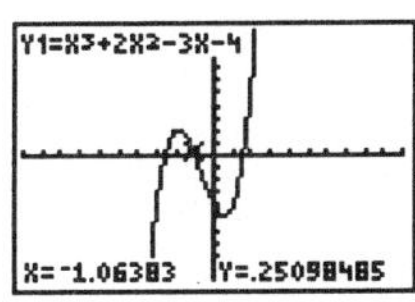

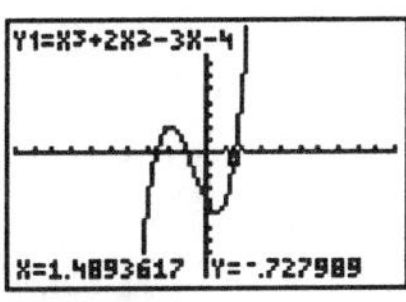

Using the TRACE feature, the zeros of this function can be estimated to be x = –2.6, –1.1, and 1.5.

Trigonometric Functions

The TI-83 is capable of calculating the sine, cosine, and tangent ratios. The inverse trigonometric functions can also be calculated. For this section, be sure that your calculator is in degree mode.

1. Press MODE. If necessary, use the DOWN ARROW to move to the Radian Degree line and highlight Degree by moving the cursor to Degree and pressing ENTER.

Example: Use trigonometric and inverse trigonometric functions to solve the right triangle.

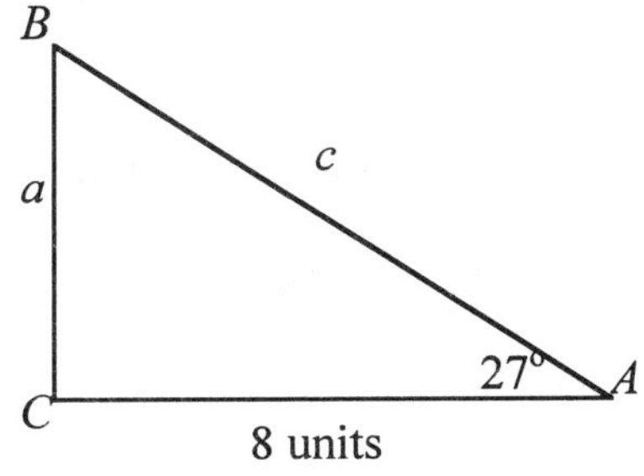

1. To determine the measure of side c, use the cosine ratio.

$$\cos 27° = \frac{8}{c}$$

$$c = \frac{8}{\cos 27°}$$

2. Press 8 ÷ COS 27 ENTER.

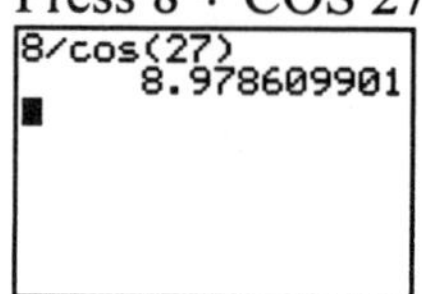

Note: The left parenthesis is automatically inserted on the TI-83.

3. To determine the length of side a, use the tangent ratio.

$$\tan 27° = \frac{a}{8}$$

4. Press 8*tan 27 ENTER.

```
8*tan(27)
          4.076203596
```

5. To determine the measure of angle B, use the inverse sine function.

$$\sin B = \frac{8}{c}$$

$$\sin B = \frac{8}{8.97}$$

$$B = \sin^{-1}\frac{8}{8.97}$$

6. Press 2nd SIN (SIN^{-1})(8 ÷ 8.97) ENTER.

NOTE: The ratio is entered in parenthesis.

```
sin⁻¹(8/8.97)
          63.10813538
```

Value

The value feature can be used to evaluate a function at a particular value of x. It works in conjunction with a graph of a function.

1. Enter an equation into Y_1.
   Example: $Y_1 = 2x^2 + 3x + 1$

2. Press ZOOM 6: Zstandard to view the function in the standard viewing window.

3. To evaluate the function at different values of x, press 2nd CALC and choose option 1: value.

4. Enter a value for x for which you would like to find the value of the function.

Example: x=2

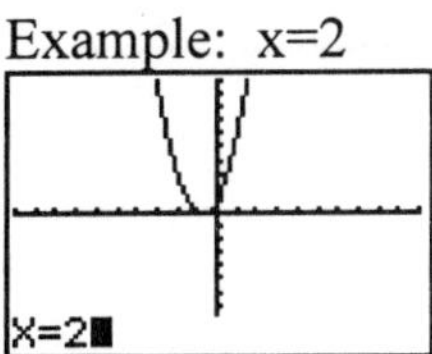

5. Press ENTER to see the value of the function at the selected value of x.

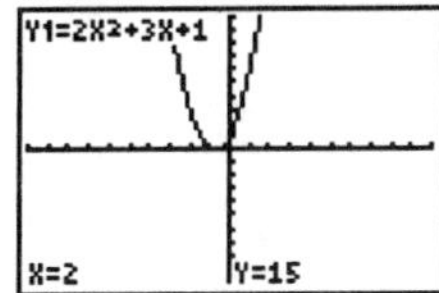

Note: You can do this many times in a row, but only for values of x in the Window Range.

$Y =$

The $Y =$ menu is used to enter functions into the calculator that will be graphed.

1. Press $Y =$.

Plot1 Plot2 Plot3
\Y1=
\Y2=
\Y3=
\Y4=
\Y5=
\Y6=
\Y7=

2. Up to 10 equations can be entered and graphed. Press the DOWN arrow to see Y_8-Y_0.

3. Type in the equation to be graphed.

Example: $Y_1 = -4x + 5$ (Note: Be sure to use the grey (-) key not the blue – key when typing the $-4x$ part of the equation.)

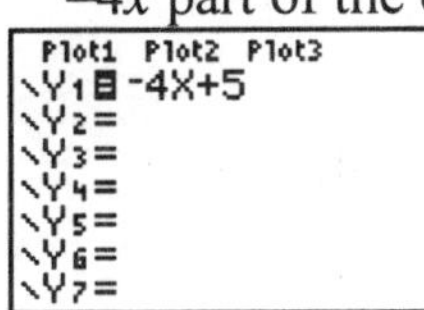

4. Press GRAPH. Depending on the viewing window, you may or may not see the graph. To view the graph in the standard viewing window, press ZOOM 6: Standard. See the ZOOM section of the manual for more details.

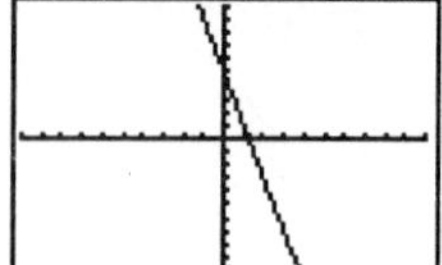

Zero

The zero function calculates the x-intercepts of a function in the graphing screen.

1. Enter a function in the $Y =$ menu.

   For example: $Y_1 = x^2 + 3x - 5$

2. Press ZOOM 6: ZStandard to view the function in the standard viewing window.

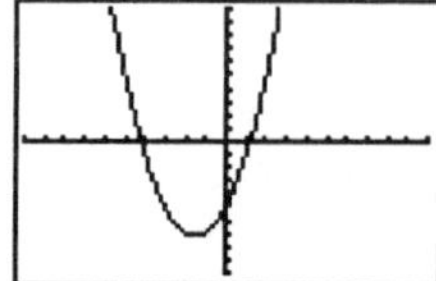

3. Press 2nd TRACE (CALC) and select option 2: zero.

4. The zero function calculates one x-intercept at a time. After selecting option 2: zero, the calculator asks for the Left Bound. This is any value to the left of the x-intercept being calculated. Use the LEFT and RIGHT arrow keys to move the cursor somewhere to the left of the zero being calculated and press ENTER.

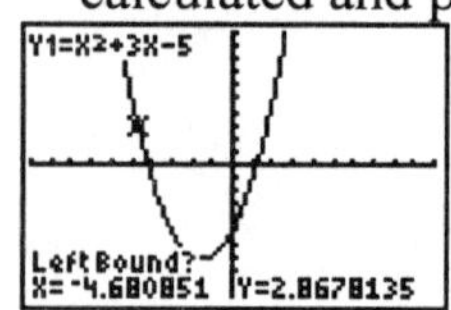

5. The calculator asks for the Right Bound. Use the LEFT and RIGHT arrow keys to move the cursor somewhere to the right of the zero being calculated and press ENTER.

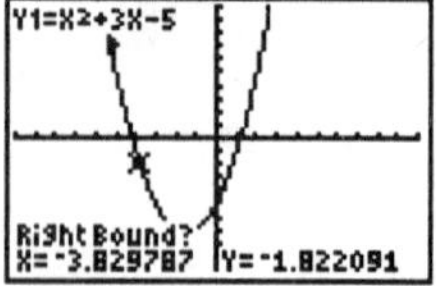

6. The calculator asks for a guess as to where the x-intercept is located. Use the LEFT and RIGH arrow keys to move the cursor as close to the x-intercept as possible and press ENTER. The x-intercept should now be displayed. Notice the y-value. It should be zero or at least very close to zero!

Y1=X²+3X-5
Guess?
X=-4.255319 Y=.34178361

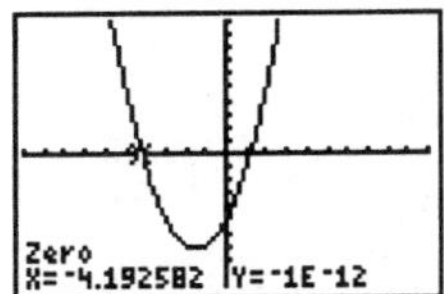

Note: In this example the y-value is -1×10^{-12}, which is extremely, close to zero! To find the other zero, repeat steps 3-6 focusing on the x-intercept on the positive x-axis.

ZOOM/WINDOW

The ZOOM button changes the viewing window rapidly with pre-set ZOOM functions.

ZOOM MEMORY
1:ZBox
2:Zoom In
3:Zoom Out
4:ZDecimal
5:ZSquare
6:ZStandard
7↓ZTrig

1. Enter an equation into the $Y =$ screen.

   Example: $Y_1 = 2x^2 + 3x + 1$.

2. Press ZOOM 6: ZStandard. ZStandard creates the standard viewing window which goes from –10 to 10 on the x-axis and –10 to 10 on the y-axis. Press WINDOW to see this.

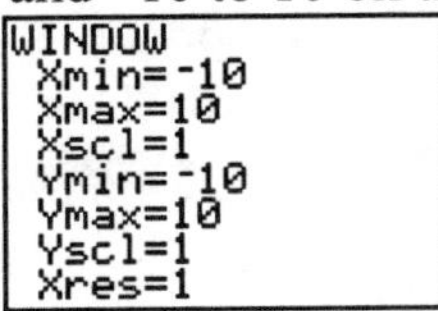
WINDOW
Xmin=-10
Xmax=10
Xscl=1
Ymin=-10
Ymax=10
Yscl=1
Xres=1

3. Press ZOOM 1: ZBox. ZBox allows zooming in on a particular region on the graph by boxing in the desired region. After pressing ZOOM 1, use the UP, DOWN, RIGHT, and/or LEFT arrow keys to move the cross hair around the screen. Think about where you would like one corner of the boxed region to be. When you have moved the cross hair to that spot, press ENTER.

For Example:

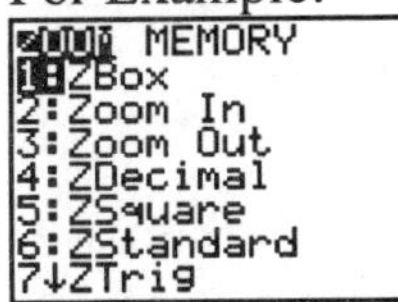
ZOOM MEMORY
1:ZBox
2:Zoom In
3:Zoom Out
4:ZDecimal
5:ZSquare
6:ZStandard
7↓ZTrig

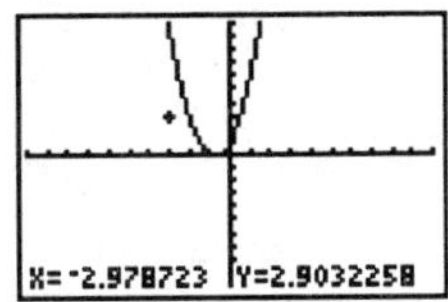

4. Use the UP, DOWN, RIGHT, and/or LEFT arrow keys to create a box around the region in which you wish to ZOOM. When the desired region is boxed in, press ENTER. When completed, press WINDOW to see how the viewing window has changed.

For Example:

X=1.7021277 Y=-.6451613

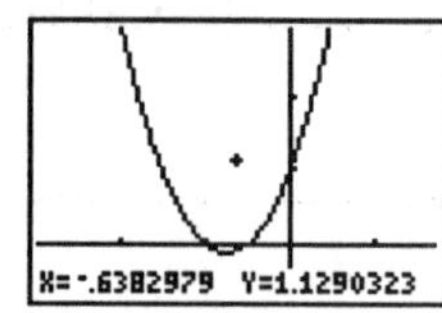

WINDOW
Xmin=-2.978723...
Xmax=1.70212766
Xscl=1
Ymin=-.64516129
Ymax=2.90322581
Yscl=1
Xres=1

5. Press ZOOM 6: ZStandard to return the graph to the standard viewing window.

6. Press ZOOM 2: Zoom In. Zoom In will do just that, centered at a point that you select. After pressing ZOOM 2: Zoom In, use the UP, DOWN, RIGHT, and/or LEFT arrow keys to move the cross hair around the screen. Stop when the desired center of the Zoom In has been located. Press ENTER to cause the Zoom In to occur. When completed, press WINDOW to see how the viewing window has changed.

For Example:

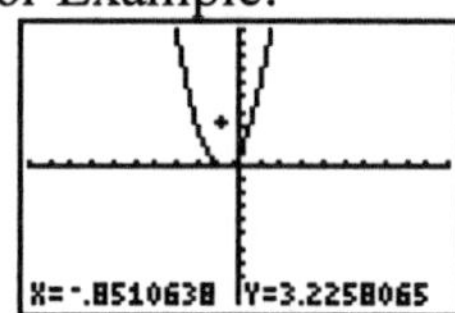

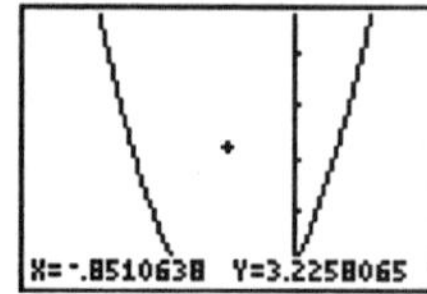

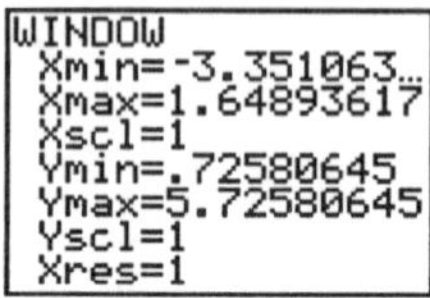

Note: The magnitude at which your calculator zooms may be different than this example.

7. Press ZOOM 6: ZStandard to return the graph to the standard viewing window.

8. Press ZOOM 3: Zoom Out. Zoom Out will do just that, centered at a point that you select. After pressing ZOOM 3: Zoom Out, use the UP, DOWN, RIGHT, and/or LEFT arrow keys to move the cross hair around the screen. Stop when the desired center of the Zoom Out has been located. Press ENTER to cause the Zoom Out to occur. When completed, press WINDOW to see how the viewing window has changed.

For Example:

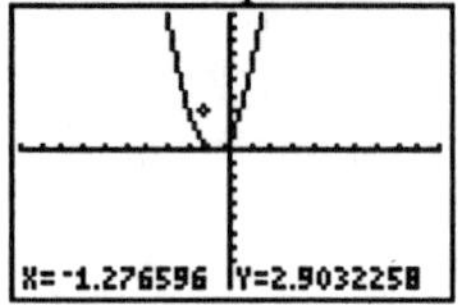

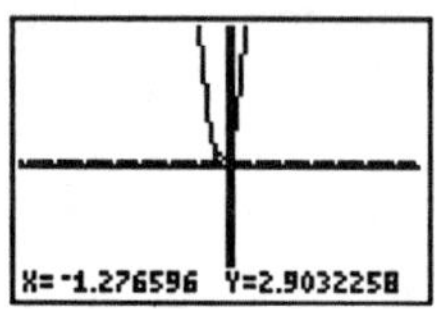

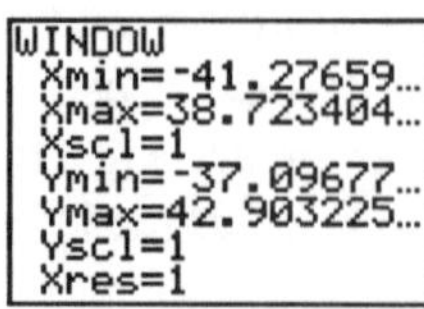

9. Press ZOOM 6: ZStandard to return the graph to the standard viewing window.

10. Press ZOOM 4: ZDecimal. ZDecimal will automatically create a viewing window which is considered "friendly." This means that when you TRACE on a graph viewed using Zdecimal, the values shown will be "friendly" decimals, showing accuracy to the nearest tenth rather than the long decimal values shown in other viewing windows. When completed, press WINDOW to see how the viewing window has changed.

For Example:

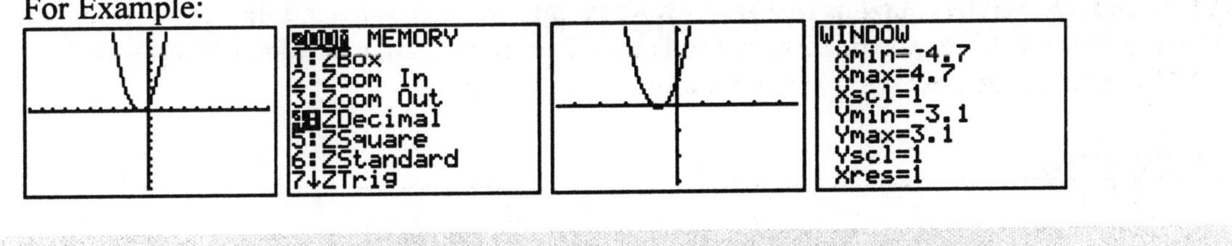
MEMORY
1:ZBox
2:Zoom In
3:Zoom Out
ZDecimal
5:ZSquare
6:ZStandard
7↓ZTrig
WINDOW
Xmin=-4.7
Xmax=4.7
Xscl=1
Ymin=-3.1
Ymax=3.1
Yscl=1
Xres=1

Calculator Keystroke Manual for the TI-85

Contents

Introduction

The graphing calculator is a tool to enable students to visualize mathematics, solve problems, think critically, and develop a conceptual understanding of the mathematics being studied. This Calculator Keystroke Manual is to provide students with a reference manual for using a graphing calculator. The intent is to allow students to focus on learning mathematics and for teachers to teach mathematics rather than the focus being on pushing buttons. It is written so that it can be referenced whenever a particular calculator function is desired. Each function, listed in alphabetical order, contains a step-by-step approach with calculator screen shots displaying what should be seen after each step. Occasionally, sample problems are given and worked through to better display that calculator function. Anytime a particular calculator button is referenced, the button name will be typed in all CAPS. If a 2nd function is referenced, the main button name is used with the actual function desired typed in parenthesis. For example, to turn the calculator OFF, type 2nd ON (OFF). If the F-keys or M-keys are referenced, the name of the function on the screen is named with the F- or M-key in parenthesis.

This manual does not provide instructions on using every function available on the calculator. With this background and subsequent mathematics courses, students will continue to develop their ability to use the calculator as an effective and efficient tool for doing mathematics.

Before beginning, be sure that the calculator is prepared correctly.

Preparing the Calculator

1. Press 2nd MORE (MODE). For the purposes of this manual, all settings on the left hand side of the screen should be highlighted. If something is not set properly, use the UP, DOWN, LEFT and RIGHT keys to move the cursor to the desired location. Press ENTER to highlight the appropriate setting.

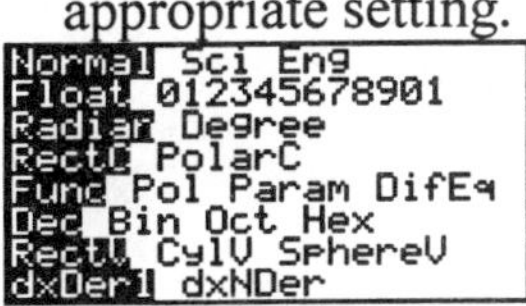

2. The HOME SCREEN is the main screen where most calculations are done.

Using This Manual

A good way to use this manual is to complete each of sections in the order given, then use the manual as a reference when needed.

1. ON/OFF
2. Contrast
3. ANS (Answer)
4. DEL/INS (Delete/Insert)
5. ENTRY
6. Square Root
7. Exponents

Absolute Value

The absolute value of a number always returns the positive value of that number. The absolute value function can be performed numerically in the home screen or can be graphed as a function in the $y=$ screen.

Example: Determine the absolute value of –5.

1. From the home screen, press 2nd × (MATH), select NUM (F1), and select abs (F5).

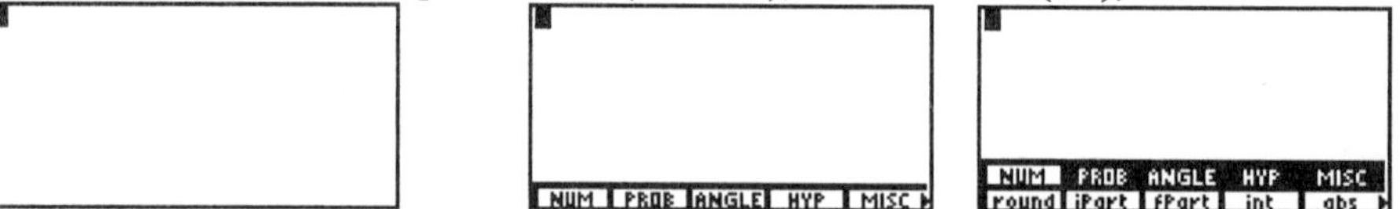

2. Enter –5 in the parentheses. Be sure to use the gray negative button not the black minus button.

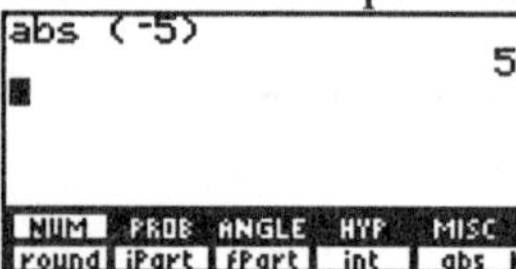

Example: Graph $y_1 = |x|$.

3. Press GRAPH and select $y=$(F1) and enter $y_1 = \text{abs}(x)$. Follow the previous instructions for accessing the abs command.

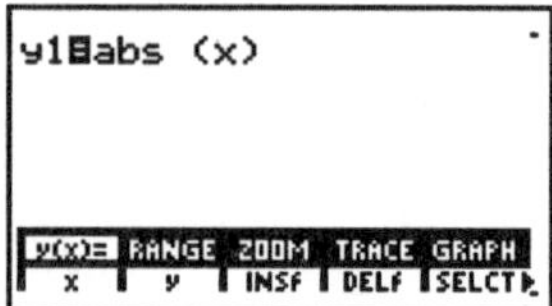

4. Press ZOOM (M3) and select ZSTD (F4) to graph the function in the standard viewing window.

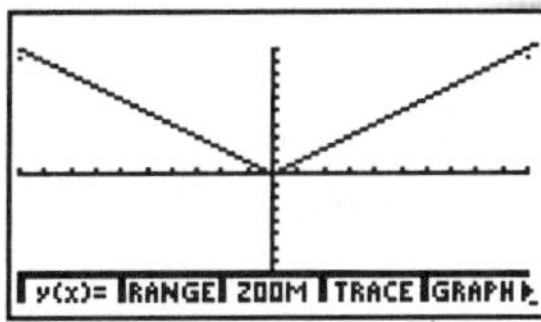

ANS

The ANS function recalls the previous answer so that additional operations can be performed on the answer to a previous operation.

Example: Start with the number 1 and continue to double it indefinitely.

1. Enter the number 1 and press ENTER. Press the multiplication button (×) then press 2 and ENTER. The ANS part of the display refers to the previous answer, in this case the 1 that was entered initially.

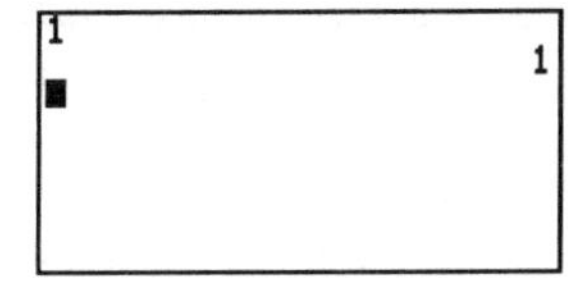

2. Continue to press ENTER until reaching the answer 64. Each time, the calculator takes the previous answer and multiples by 2.

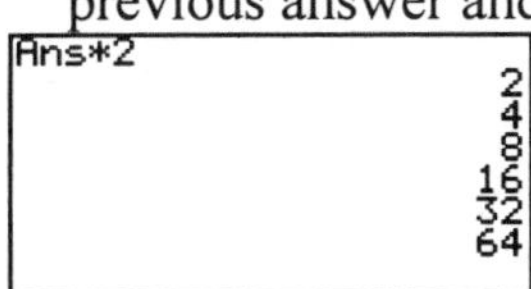

The ANS function can be accessed by typing 2nd (-) (ANS).

3. Press 7, press +, press 2nd (-) (ANS), then ENTER. The calculator takes the last answer (64), adds 7, and returns the answer of 71.

Box and Whisker Plots

Unlike the TI-82, 83, and 86, the TI-85 does not have the Box and Whisker Plot option.

Contrast

The contrast for the calculator screen can be adjusted to be darker or lighter. As your batteries become older, it will become necessary to adjust the contrast so that you can more easily read the calculator screen.

1. Press 2nd and release.

2. Hold the DOWN ARROW key to make the screen lighter. Hold the UP ARROW key to make the screen darker. Each time the DOWN or UP ARROW key is released, the 2nd key must be pressed again to continue to adjust the contrast.

3. As the UP or DOWN ARROW keys are held, notice the number in the upper right hand corner. The contrast is measured from 1 – 9 where 1 is the lightest setting and 9 is the darkest. If the contrast needs to be 9 in order for you to see the screen, you probably need to replace the batteries.

DEL/INS

The DEL (Delete) key deletes the character under the current cursor location. The 2nd DEL (INS) (Insert) key inserts a character to the left of the current cursor location.

Example: You desire to type 9+6–5*34, but accidentally type 9+36–5*4.

1. Enter 9+36–5*4. Use the LEFT arrow to move the cursor to the 3 location.

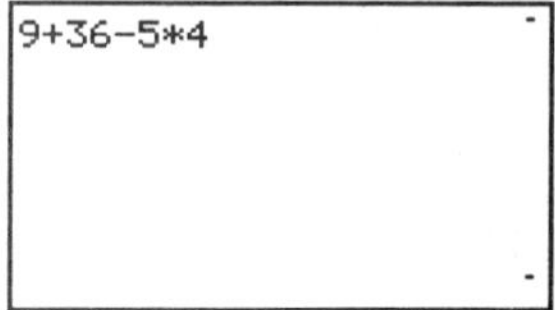

2. Press DEL. This deletes the unneeded 3.

3. Use the RIGHT ARROW to move the cursor to the 4 location. Press 2nd DEL (INS) and enter a 3. This inserts the 3 to the left of the 4 as desired. Notice that the cursor changed after pressing 2nd DEL (INS) so that you knew that the calculator was in the insert mode.

9+6-5*_

9+6-5*34■

ENTRY

The ENTRY function returns previously entered operations on the calculator screen so that you may edit them.

1. Press 2nd ENTER (ENTRY). The last operation entered into the computer appears on the screen and the cursor is at the right end of the line. Use the UP, DOWN, RIGHT, LEFT arrow keys to edit the line.

2. Press 2nd ENTER (ENTRY) again. The TI-85 returns only the last entry made.

Error Messages

Occasionally, error messages are given to indicate some problem with the function you are asking the calculator to perform. Three common error messages and how to correct them are provided.

Syntax Error

Syntax refers to the way in which the function or command was entered. One common error deals with use of parentheses.

Example: You intend to enter 6*(3+2) but accidentally enter 6*3+2).

1. In the Home Screen, enter 6*3+2) and press ENTER.

6*3+2)

ERROR 07 SYNTAX

GOTO QUIT

2. Because of the missing opening parenthesis, the calculator returns a syntax error. Notice the options at the bottom of the screen. You can QUIT and start over, or you can GOTO the error so that it can be corrected. Select GOTO (F1).

3. To correct the mistake, insert a parenthesis before the 3 using INS. Press ENTER.

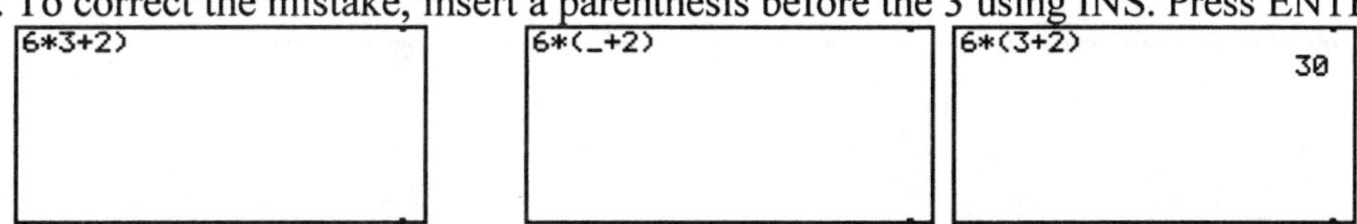

Graph Range Error

The Graph Range Error refers to an incorrect RANGE entry.

1. Enter an equation into $y =$.

For example: $y_1 = x^3 + 3$

2. Press ZOOM (M3) and select ZSTD (F4) to graph the function in the standard viewing window.

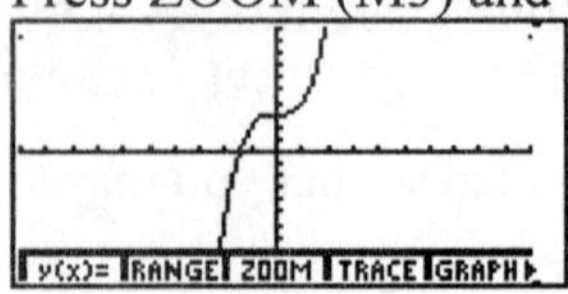

3. Press RANGE (F2). Imagine that you incorrectly changed the RANGE as shown below. Press GRAPH (F5). A Graph Range Error occurs.

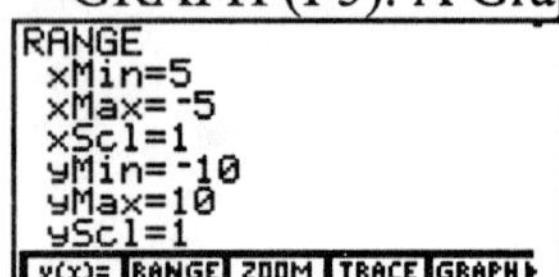

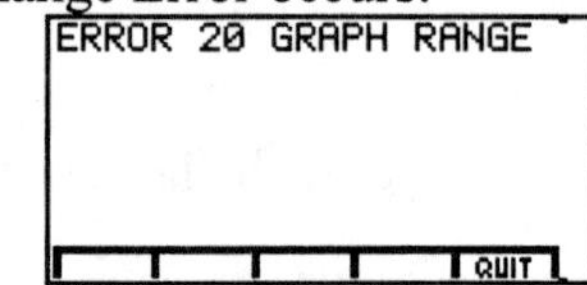

4. The only option is to QUIT (F5). To correct the error, press GRAPH, select RANGE (F2) and change the xMin and xMax as shown. Press GRAPH (F5).

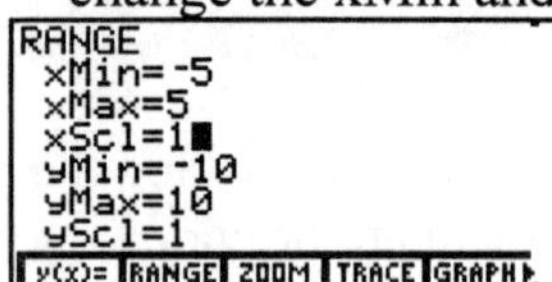

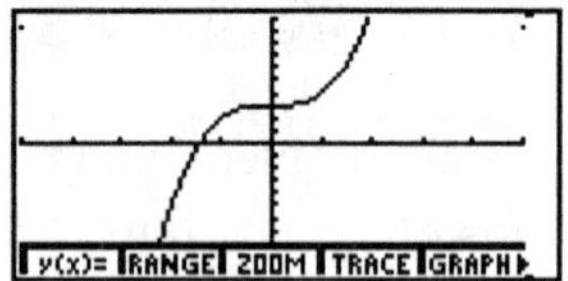

EXIT

The EXIT key will back out of one screen towards the home screen.
Example: Press GRAPH then ZOOM (F3). To get back to the home screen, you can press EXIT to return to the GRAPH menu, then EXIT again to return to the home screen.

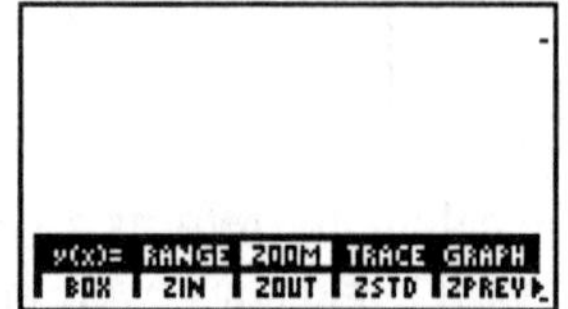

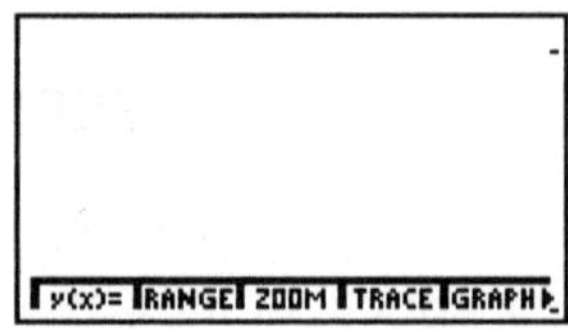

Exponents

Exponents of 2 can be entered into the calculator 2 different ways. Exponents greater than 2 are entered using the caret key.

Example: Evaluate $3^2 + 7^2 - 8^4 + 2^7$.

1. To enter the powers of 2, use the x^2 key that is located in the middle of the first column of keys. To enter powers greater than 2, use the caret key (∧) which is located below the CLEAR key in the last column.

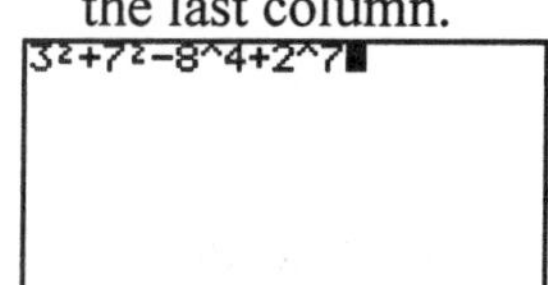

2. Press ENTER.

```
3²+7²-8^4+2^7
                -3910
```

GRAPH and $y =$

The GRAPH and $y =$ menu is used to enter functions into the calculator that will be graphed.

1. Press GRAPH and $y =$ (F1).

```
y1=

y(x)= RANGE ZOOM TRACE GRAPH
x  y  INSf  DELf  SELCT
```

2. Up to 99 equations can be entered and graphed. After entering an equation, press ENTER to enter another equation.

3. Type in the equation to be graphed.

   Example: $y_1 = -4x + 5$ (Note: Be sure to use the gray (-) key not the black – key when typing the $-4x$ part of the equation.)

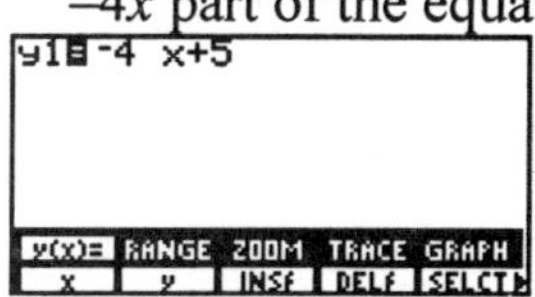

4. Press GRAPH (M5). Depending on the viewing window, you may or may not see the graph. To view the graph in the standard viewing window, press ZOOM (F3) and select ZSTD (F4). See the ZOOM section of the manual for more details.

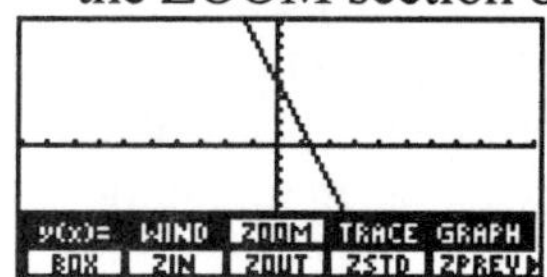

5. Press CLEAR to remove the menu choices from the bottom of the screen.

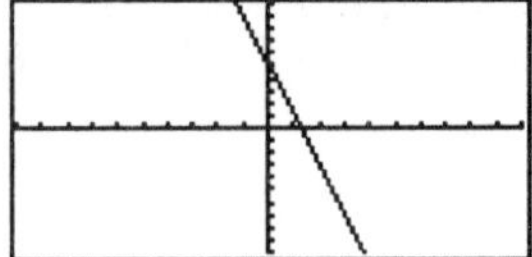

Histograms

Histograms are a way to display and analyze data. Create a histogram for the data given.

The shoe size of some the children in the class are shown.

Child	1	2	3	4	5	6	7	8	9
Shoe Size	5	5	4	5	4	3	6	5	4

1. Clear any equations from the y = menu.

2. Press STAT and select EDIT (F2).

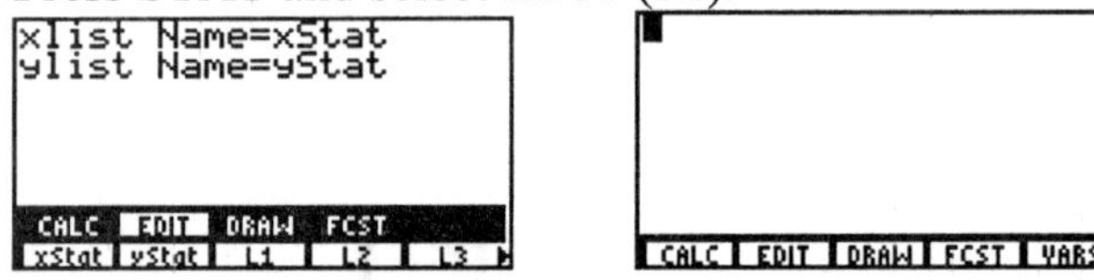

You can name your data in this screen, but it is usually sufficient to leave the names as they are (xStat and yStat). Press ENTER twice to move through this screen and into the data entry screen. If data has been entered previously, it can be cleared by pressing CLRxy (F5).

To enter the data, type in the shoe sizes only for the x-coordinate. For this example, leave the y-coordinate as 1 in each case. Press ENTER after each entry.

2. Adjust your RANGE (press GRAPH and select RANGE (F2)) so that it reflects the values entered into the lists. The yMin should be set at 0 because 0 is the least number of times a particular shoe size is seen. yMax should be set at 5 since the largest number of times a particular shoe size is seen is 4 (4 children wear size 5 shoes).

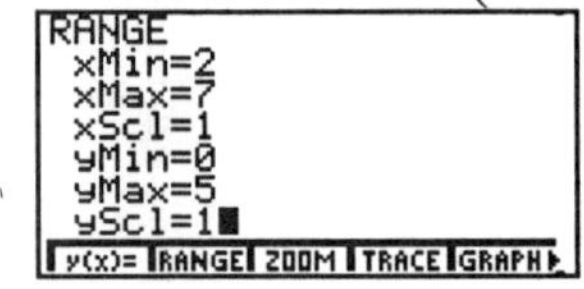

3. To view the histogram, STAT, DRAW (F3), and HIST (F1). Press CLEAR to view the entire histogram.

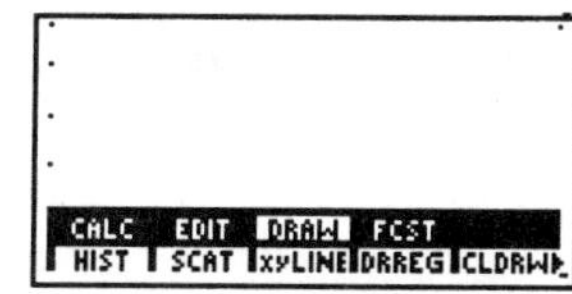

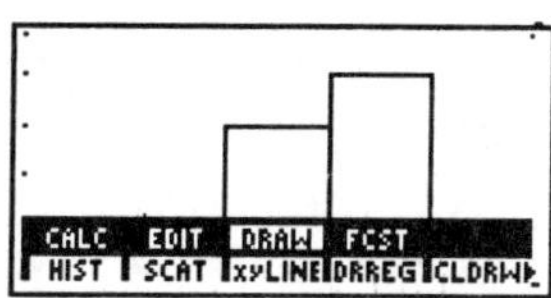

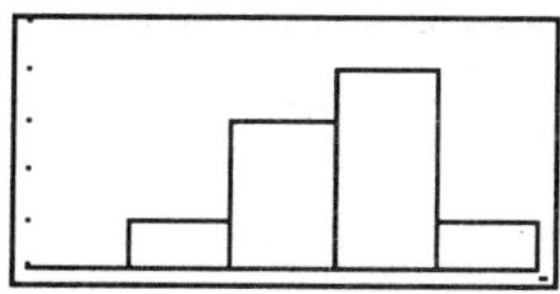

ISECT

The ISECT (intersect) function will calculate an intersection point of two graphed equations.

1. Enter the equations in y_1 and y_2.
   Example: $y_1 = 2x^2 + 3x + 1$
   $y_2 = 3x + 4$

2. Select an appropriate viewing window. In this case the standard viewing window may be used (ZSTD).

3. Either press GRAPH (M5) or ZOOM (M3) and select ZSTD (F4).

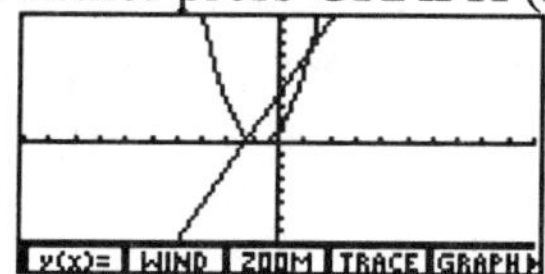

4. To determine the first quadrant point of intersection for these two graphs, press MORE, MATH (F1), MORE, ISECT (F5).

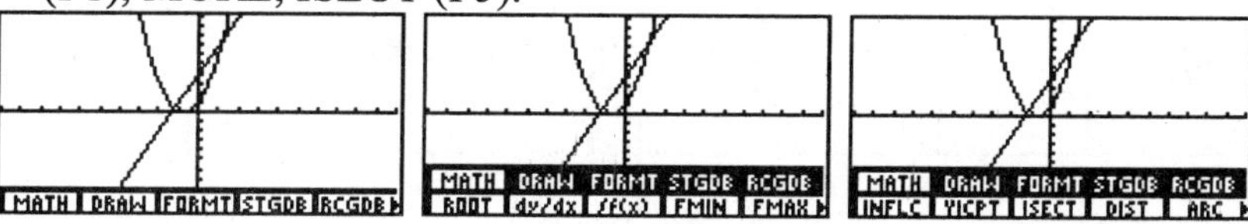

5. Automatically returning to the graph screen, notice the 1 in the upper right hand corner of the screen. This represents the calculator asking for the first curve. The first curve is the function defined in y_1. Move the cursor to the approximate point of intersection using the LEFT or RIGHT arrow keys and then press ENTER.

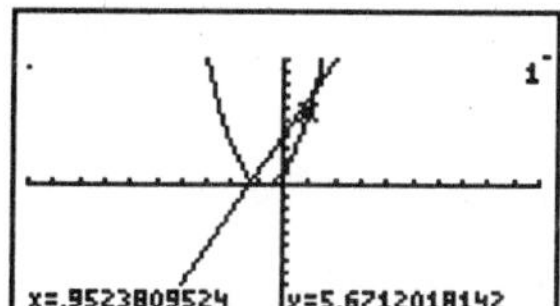

6. Notice the 2 in the upper right hand corner of the screen. This represents the calculator asking for the second curve. The second curve is the function defined in y_2. Since the cursor is already near the point of intersection, simply press ENTER. This causes the calculator to determine the point of intersection.

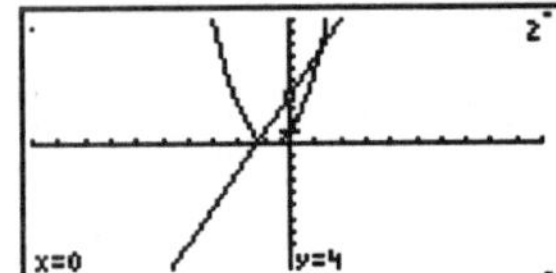

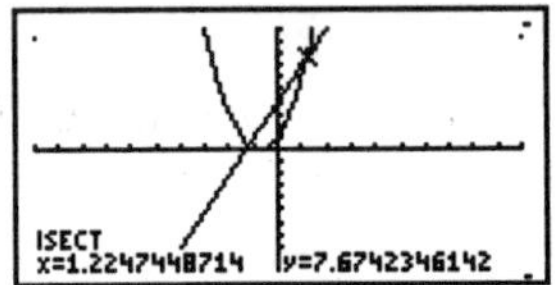

ON/OFF

1. Press ON to turn the calculator on.

2. To turn the calculator off, press 2nd ON (OFF).

After several minutes of inactivity, the calculator will automatically shut off.

One-Variable Statistics

The 1-Variable Statistics command returns the following statistics: mean ($\bar{x}$), sum of x ($\sum x$), sum of the squares of x ($\sum x^2$), sample standard deviation (Sx), population standard deviation (σx), and the amount of data entered (n).

Example: Determine all of the one-variable statistics for the following data set.

The ACT scores for some students in a particular school is given.

Student	1	2	3	4	5	6	7	8	9
ACT Score	20	22	30	32	28	21	20	34	18

1. Clear any equations from the $y =$ menu.

2. Press STAT and select EDIT (F2).

You can name your data in this screen, but it is usually sufficient to leave the names as they are (xStat and yStat). Press ENTER twice to move through this screen and into the data entry screen.

If data has been entered previously, it can be cleared by pressing CLRxy (F5).

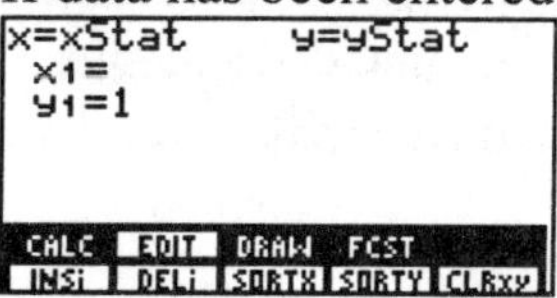

To enter the data, type in the ACT Score only for the x coordinate. For this example, leave the y coordinate as 1 in each case. Press ENTER after each entry.

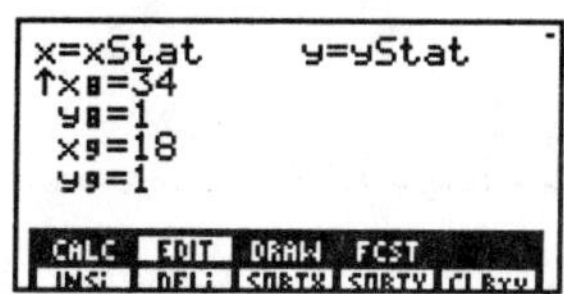

2. From the home screen (press 2nd EXIT (QUIT)), press STAT and select CALC (F1). Press ENTER twice to select the xStat and yStat data sets. Select 1-VAR (F1). The statistics will be displayed on the screen.

QUIT

This returns the calculator to the home screen from any other menu.

Example: Imagine that you are in the WINDOW screen and need to get back to the home screen.

1. Press WINDOW. To go back to the home screen, press 2nd EXIT (QUIT).

Regression

The regression feature creates a mathematical model for data that has been entered in the Lists. For example, create a scatter plot and calculate a mathematical model for the set of data given.

The revenue for Dell Computer Corporation from 1985 through 1993 is listed in the table. The revenue is given in millions of dollars.

Year	1985	1986	1987	1988	1989	1990	1991	1992	1993
Revenue	33.7	69.5	159.0	257.8	388.6	546.2	889.0	2013.9	2873.2

(Source: Dell Computer Corporation)

1. Clear any equations from the $y =$ menu.

2. Press STAT and select EDIT (F2).

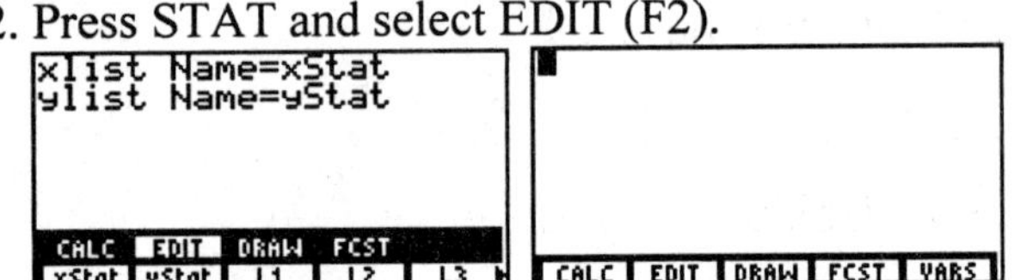

You can name your data in this screen, but it is usually sufficient to leave the names as they are (xStat and yStat). Press ENTER twice to move through this screen and into the data entry screen. If data has been entered previously, it can be cleared by pressing CLRxy (F5).

To enter the data, type in the ordered pairs (x, y), which in this case is the year and the revenue. Press ENTER after each entry. For this example, enter only the last two digits of the year.

3. Adjust your RANGE by pressing GRAPH and RANGE (F2) so that it reflects the values entered into your lists.

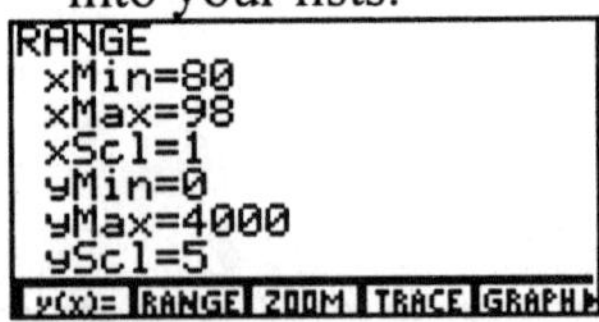

4. Set up the plot by pressing STAT again and selecting the DRAW (F3). To create a scatter plot, choose SCAT (F2).

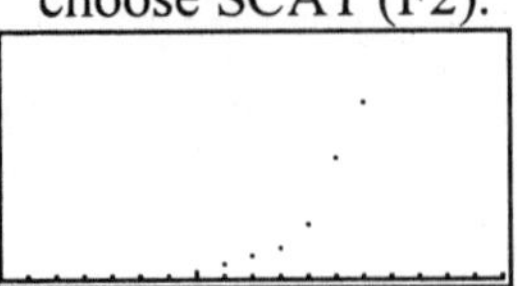

5. To have the calculator create an equation to model your data, press STAT again. This time select CALC (F1). Press enter twice to move through the xlist, ylist name screen. Then, at the bottom of the screen, many options appear for the type of regression that you choose to perform.

For example: Select EXPR (F5) (Exponential Regression). The result are the values of a and b which fit the form $y = ab^x$.

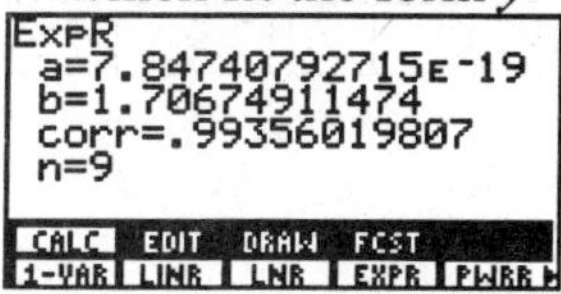

You can record this information on your paper and enter the equation in y =, or...

6. To graph the function without needing to remember or write down the regression model, press GRAPH and select y = (F1). Press 2nd 3 (VARS), press MORE, MORE, and select STAT (F3).

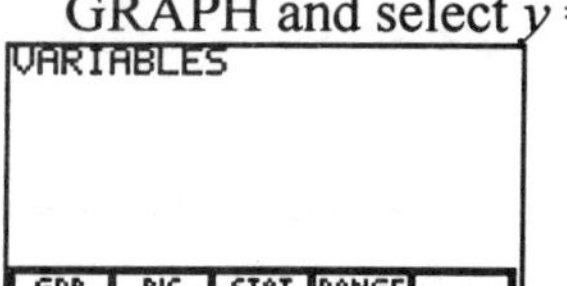

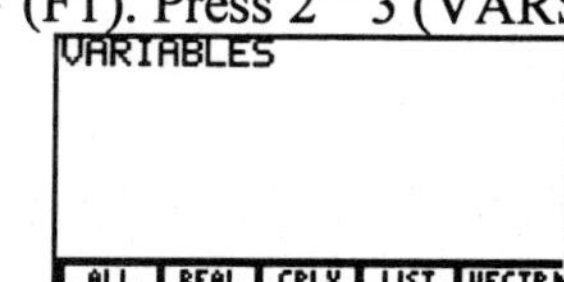

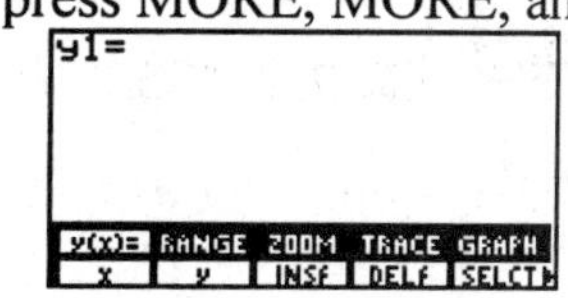

7. Use the DOWN arrow to select the RegEq option. Press ENTER. This pastes the RegEq command into the y = menu.

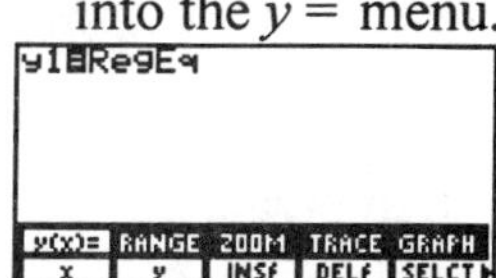

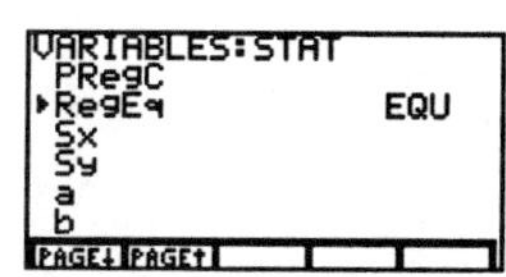

8. Press GRAPH (M5). Press STAT, DRAW (F3), SCAT (F2) to view the scatter plot again. Press CLEAR to view the entire graph.

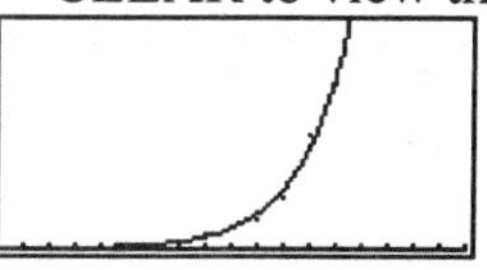

Root

The Root function calculates the x-intercepts of a function in the graphing screen.

1. Enter a function in the GRAPH, y = menu.
   For example: $y_1 = x^2 + 3x - 5$

2. Press ZOOM (F3) and select ZSTD (F4) to view the function in the standard viewing window. Press CLEAR to remove the menu choices from the bottom of the screen.

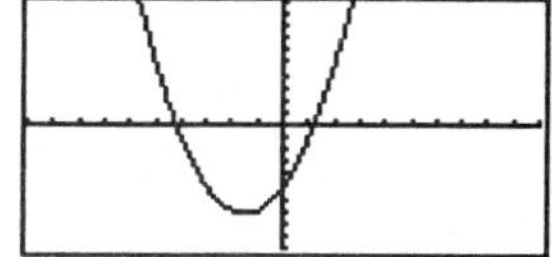

3. Press GRAPH then MORE and select MATH (F1). Select ROOT (F3).

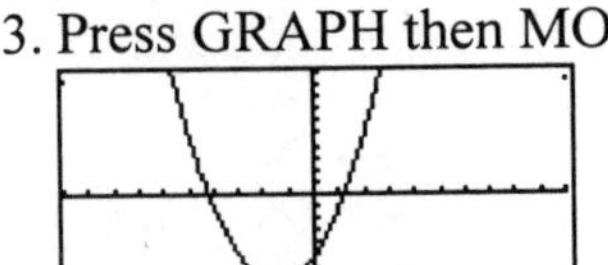

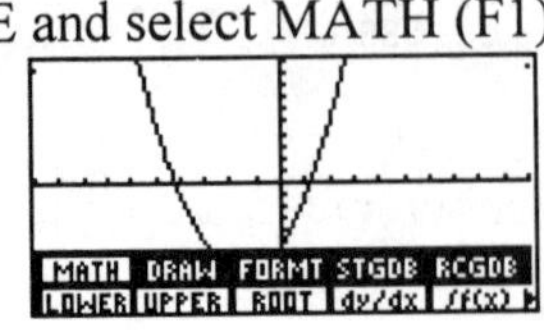

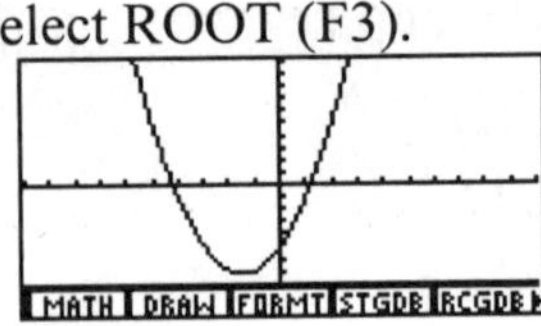

4. The root function calculates one x-intercept at a time. Use the LEFT and RIGHT arrow keys to move the cursor somewhere near the root being calculated and press ENTER.

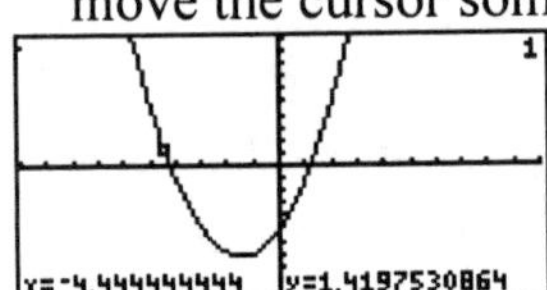

5. The calculator finds the x-intercept (root) and displays it at the bottom of the screen. Notice the y-value. It should be zero or at least very close to zero!

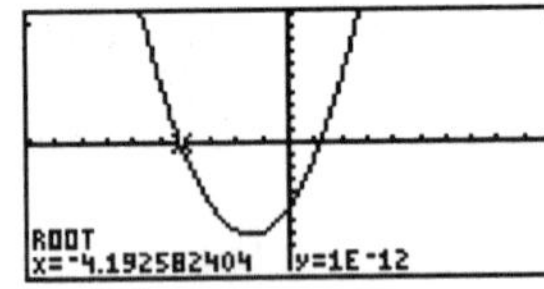

Note: In some cases the y-value could be -1×10^{-12}, for example, which is extremely, close to zero!

6. To find the other root, repeat steps 3-6 focusing on the x-intercept on the positive x-axis.

Scatter Plots - See Regression

Scientific Notation

The calculator will automatically go in to scientific notation when the result of an operation is either very small or very large.

Example: It is estimated that a satellite orbiting the earth has traveled 10,000,000 miles. How many centimeters is that?

1. Enter 10,000,000*5280*12*2.54 then press ENTER. (Note: There are 5280 feet in one mile, 12 inches in one foot, and 2.54 centimeters in one inch.)

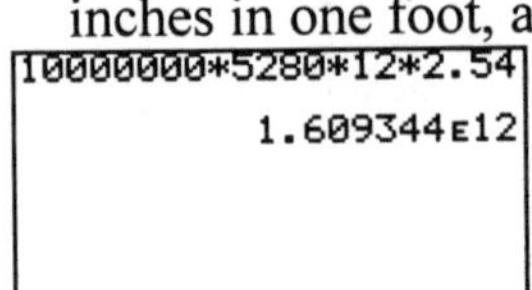

2. Notice the notation used by the calculator. The 1.609344E12 means 1.609344×10^{12}, or 1,609,344,000,000 centimeters.

Square Root

The calculator will calculate thc square root of a number and return the positive root, accurate up to 11 decimal places.

Example: What is the square root of 111?

1. Enter the square root function first. Press 2nd x^2 ($\sqrt{\ }$). Type then number that you which to calculate the square root, in this 111. Press ENTER.

```
√111
        10.5356537529
```

Note: Parentheses are needed if you need to calculate the square root of the sum or difference of numbers.

Example: What is the square root of 23 + 79?

2. Press 2nd x^2 ($\sqrt{\ }$). Press the left parenthesis. Enter 23+79 and close the parentheses. Press ENTER.

```
√(23+79)
        10.0995049384
```

Tables

Unlike the TI-82, 83, and 86, the TI-85 does not have a built-in table feature. There are programs available to create tables on the TI-85.

TRACE

The TRACE feature allows you to trace along the graph of a function and view the coordinates of the pixels at the bottom of the screen.

Example: Graph the function $y = x^3 + 2x^2 - 3x - 4$ and use the TRACE feature to approximate the x-intercepts of the graph.

1. Enter the equation into y_1 using the $y =$ screen (press GRAPH and select $y =$ (F1)). Press ZOOM (M3) and select ZSTD (F4) to graph the function in the standard viewing window.

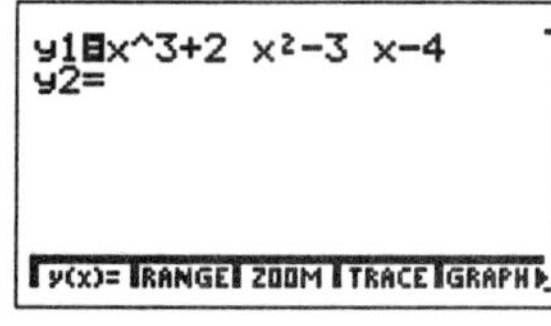

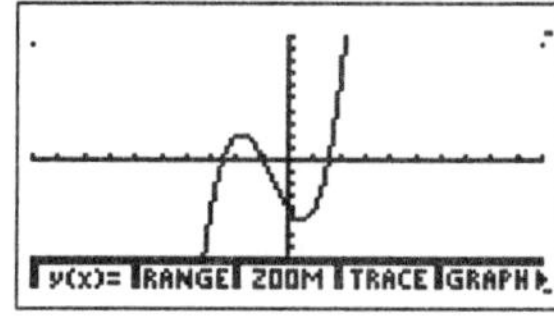

2. Press TRACE (F4). Use the LEFT and RIGHT ARROW keys to move the cursor along the graph of the curve. Estimate the zeros of the function by moving the cursor as close to the x-intercepts as possible.

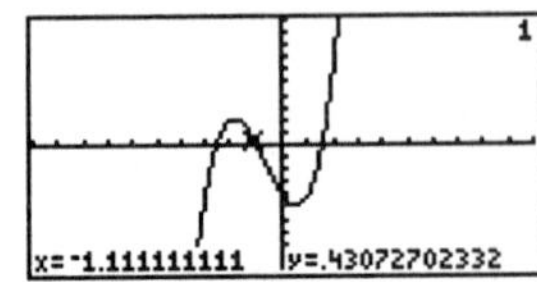

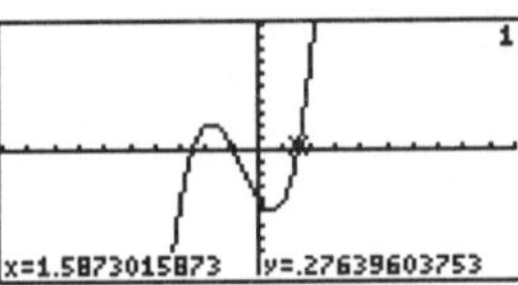

Using the TRACE feature, the zeros of this function can be estimated to be x = –2.6, –1.1, and 1.6.

Trigonometric Functions

The TI-85 is capable of calculating the sine, cosine, and tangent ratios. The inverse trigonometric functions can also be calculated. For this section, be sure that your calculator is in degree mode.

1. Press 2nd MORE (MODE). If necessary, use the DOWN ARROW to move to the Radian Degree line and highlight Degree by moving the cursor to Degree and pressing ENTER.

Example: Use trigonometric and inverse trigonometric functions to solve the right triangle.

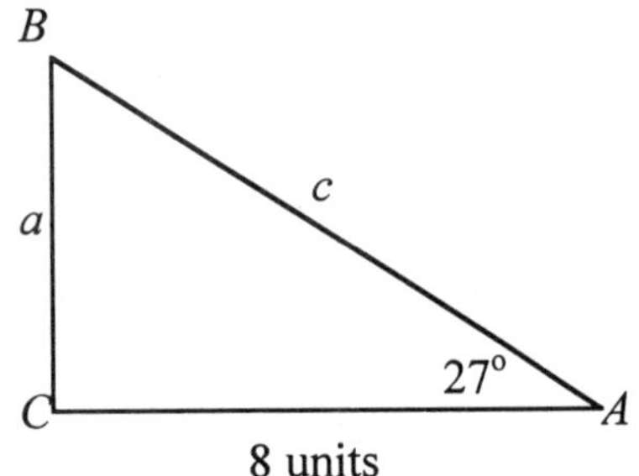

1. To determine the measure of side c, use the cosine ratio.

$$\cos 27° = \frac{8}{c}$$

$$c = \frac{8}{\cos 27°}$$

2. Press 8 ÷ COS 27 ENTER.

```
8/cos (27)
              8.97860990107
```

Note: The left parenthesis is not automatically inserted on the TI-85.

3. To determine the length of side a, use the tangent ratio.

$$\tan 27° = \frac{a}{8}$$

4. Press 8*tan 27 ENTER.

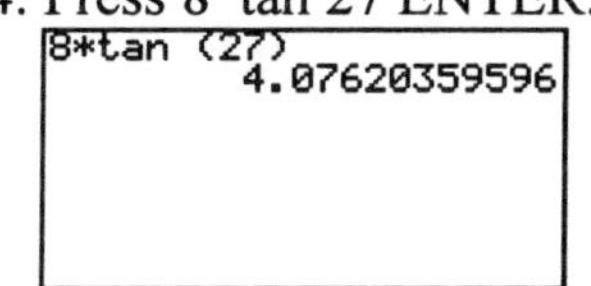

5. To determine the measure of angle B, use the inverse sine function.

$$\sin B = \frac{8}{c}$$

$$\sin B = \frac{8}{8.97}$$

$$B = \sin^{-1}\frac{8}{8.97}$$

6. Press 2nd SIN (SIN^{-1})(8 ÷ 8.97) ENTER.

NOTE: The ratio is entered in parenthesis.

Value

The value feature can be used to evaluate a function at a particular value of x. It works in conjunction with a graph of a function.

1. Enter an equation into y_1.

   Example: $y_1 = 2x^2 + 3x + 1$

2. Press ZOOM (M3) and ZSTD (F4) to view the function in the standard viewing window.

3. To evaluate the function at different values of x, press MORE twice and select EVAL (F1).

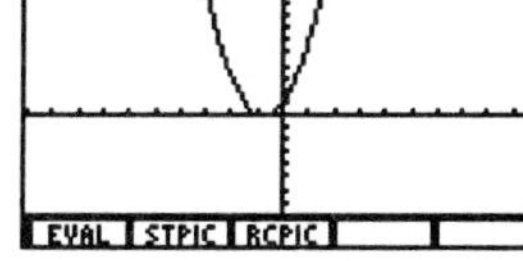

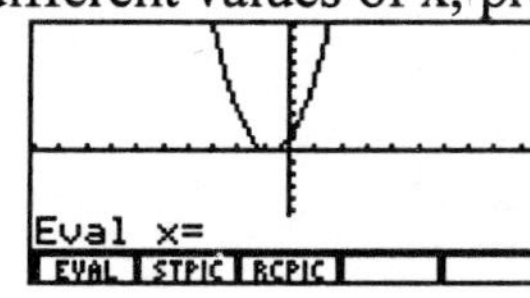

4. Enter a value for x for which you would like to find the value of the function.

   Example: x=2

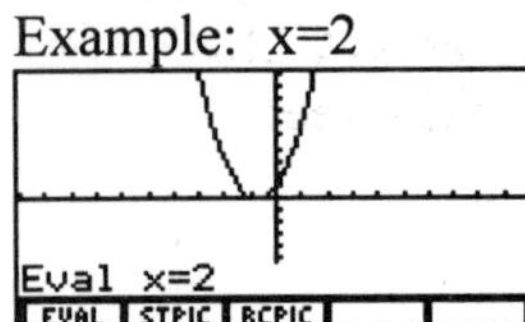

5. Press ENTER to see the value of the function at the selected value of x.

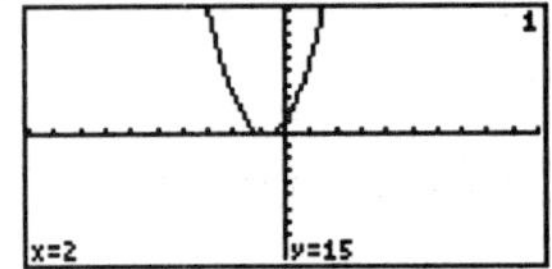

   To evaluate the function at another value of x, press GRAPH and repeat steps 3-5.

ZOOM/RANGE

The ZOOM button changes the viewing window rapidly with pre-set ZOOM functions.

1. Enter an equation into the y = screen.

   Example: $y_1 = 2x^2 + 3x + 1$.

2. Press ZOOM (M3) and select ZSTD (F4). ZSTD creates the standard viewing window which goes from –10 to 10 on the x-axis and –10 to 10 on the y-axis. Press RANGE (F2) to see this.

3. Press ZOOM (F3) and select BOX (F1). BOX allows zooming in on a particular region on the graph by boxing in the desired region. After pressing BOX (F1), use the UP, DOWN, RIGHT, and/or LEFT arrow keys to move the cross hair around the screen. Think about where you would like one corner of the boxed region to be. When you have moved the cross hair to that spot, press ENTER.

   For Example:

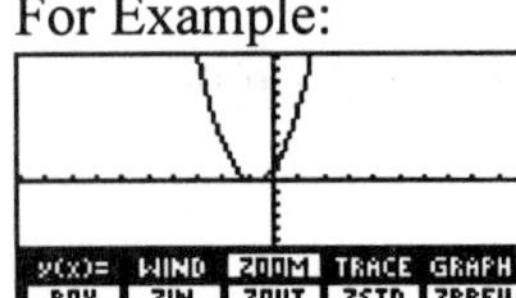

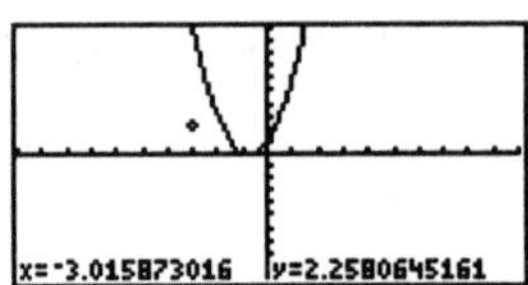

4. Use the UP, DOWN, RIGHT, and/or LEFT arrow keys to create a box around the region in which you wish to ZOOM. When the desired region is boxed in, press ENTER. When completed, press RANGE (F2) to see how the viewing window has changed. You can press CLEAR after the graph has been re-drawn to remove the menu options at the bottom of the screen.

For Example:

```
RANGE
 xMin=-3.01587301587
 xMax=1.4285714286
 xScl=1
 yMin=-.64516129032
 yMax=2.2580645161
 yScl=1
 y(x)= RANGE ZOOM TRACE GRAPH
```

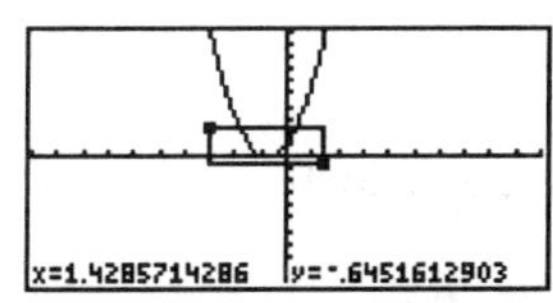

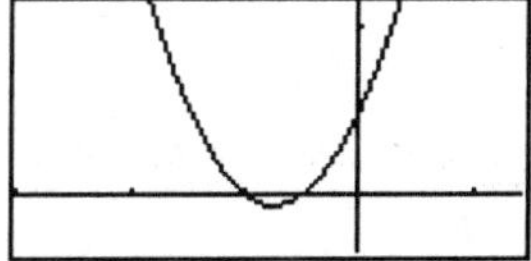

5. Press ZOOM (F2) and select ZSTD (F4) to return the graph to the standard viewing window.

6. Press ZOOM (F3) and select ZIN (F2). ZIN will do zoom in, centered at a point that you select. After pressing ZIN (F2), use the UP, DOWN, RIGHT, and/or LEFT arrow keys to move the cross hair around the screen. Stop when the desired center of the ZIN has been located. Press ENTER to cause the Zoom In to occur. When completed, press GRAPH and RANGE (F2) to see how the viewing window has changed.

For Example:

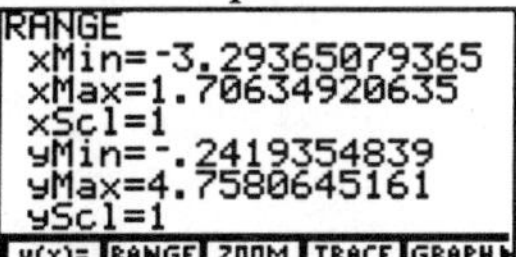

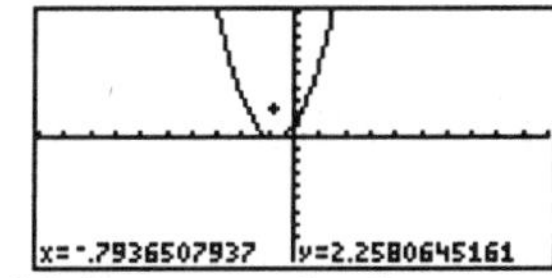

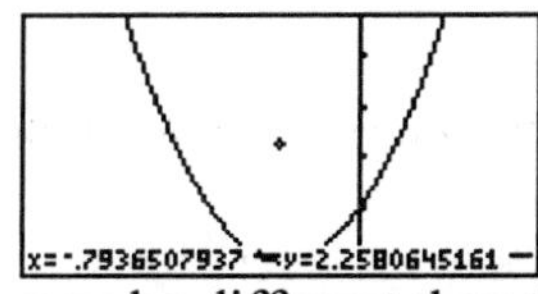

Note: The magnitude at which your calculator zooms may be different than this example. If you continue to press ENTER, the calculator will continue to Zoom In.

7. Press ZOOM (F3) and select ZSTD (F4) to return the graph to the standard viewing window.

8. Press ZOOM (F3) and select ZOUT (F3). ZOUT will zoom out, centered at a point that you select. After pressing ZOUT (F3), use the UP, DOWN, RIGHT, and/or LEFT arrow keys to move the cross hair around the screen. Stop when the desired center of the ZOUT has been located. Press ENTER to cause the Zoom Out to occur. When completed, press GRAPH and RANGE (F2) to see how the viewing window has changed.

For Example:

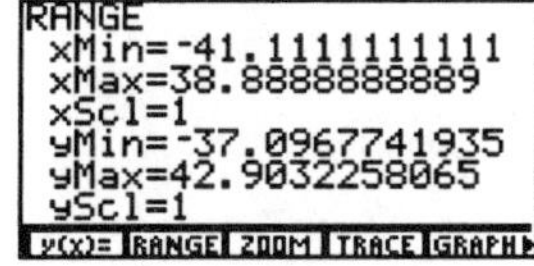

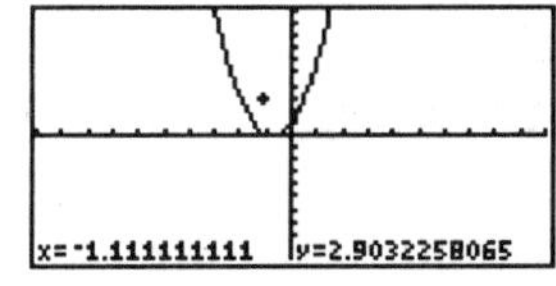

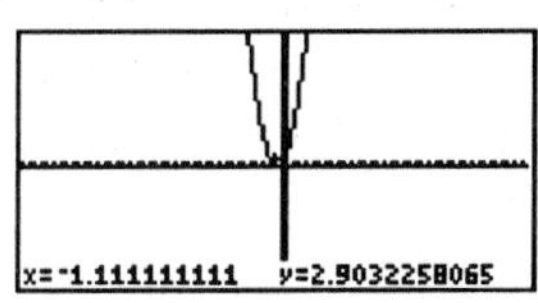

9. Press ZOOM (F3) and select ZSTD (F4) to return the graph to the standard viewing window.

10. Press ZOOM (F3) then MORE and select ZDECM (F4). ZDECM will automatically create a viewing window which is considered "friendly". This means that when you TRACE on a graph viewed using ZDECM, the values shown will be "friendly" decimals, showing accuracy to the nearest tenth rather than the long decimal values shown in other viewing windows. When completed, press RANGE (F2) to see how the viewing window has changed.

For Example:

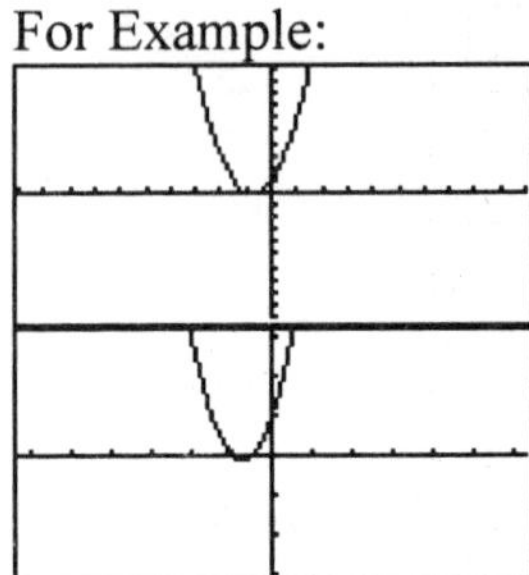

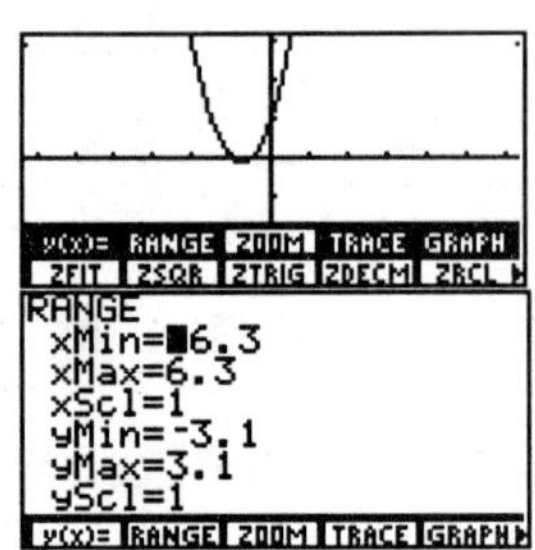

Calculator Keystroke Manual for the TI-86

Contents

Introduction

The graphing calculator is a tool to enable students to visualize mathematics, solve problems, think critically, and develop a conceptual understanding of the mathematics being studied. This Calculator Keystroke Manual is to provide students with a reference manual for using a graphing calculator. The intent is to allow students to focus on learning mathematics and for teachers to teach mathematics rather than the focus being on pushing buttons. It is written so that it can be referenced whenever a particular calculator function is desired. Each function, listed in alphabetical order, contains a step-by-step approach with calculator screen shots displaying what should be seen after each step. Occasionally, sample problems are given and worked through to better display that calculator function. Anytime a particular calculator button is referenced, the button name will be typed in all CAPS. If a 2nd function is referenced, the main button name is used with the actual function desired typed in parenthesis. For example, to turn the calculator OFF, type 2nd ON (OFF). If the F-keys or M-keys are referenced, the name of the function on the screen is named with the F- or M-key in parenthesis.

This manual does not provide instructions on using every function available on the calculator. With this background and subsequent mathematics courses, students will continue to develop their ability to use the calculator as an effective and efficient tool for doing mathematics.

Before beginning, be sure that the calculator is prepared correctly.

Preparing the Calculator

1. Press 2nd MORE (MODE). For the purposes of this manual, all settings on the left hand side of the screen should be highlighted. If something is not set properly, use the UP, DOWN, LEFT and RIGHT keys to move the cursor to the desired location. Press ENTER to highlight the appropriate setting.

2. Press 2nd + (STAT). Press PLOT (F3). Press PlOff (F5). Press ENTER.

3. The HOME SCREEN is the main screen where most calculations are done.

Using This Manual

A good way to use this manual is to complete each of sections in the order given, then use the manual as a reference when needed.

1. ON/OFF
2. Contrast
3. ANS (Answer)
4. DEL/INS (Delete/Insert)
5. ENTRY
6. Square Root
7. Exponents
8. Scientific Notation
9. QUIT
10. EXIT
11. Fractions
12. Graph and $Y=$
13. ZOOM/WINDOW
14. TRACE
15. Tables
16. Absolute Value
17. Value
18. Evaluate a Function
19. ISECT (Intersect)
20. Root
21. One-Variable Statistics
22. Histograms
23. Box and Whisker Plots
24. Regression/Scatter Plots/Lists
25. Error Messages
26. Trigonometric Functions

Absolute Value

The absolute value of a number always returns the positive value of that number. The absolute value function can be performed numerically in the home screen or can be graphed as a function in the $y =$ screen.

Example: Determine the absolute value of –5.

1. From the home screen, press 2nd × (MATH), select NUM (F1), and select abs (F5).

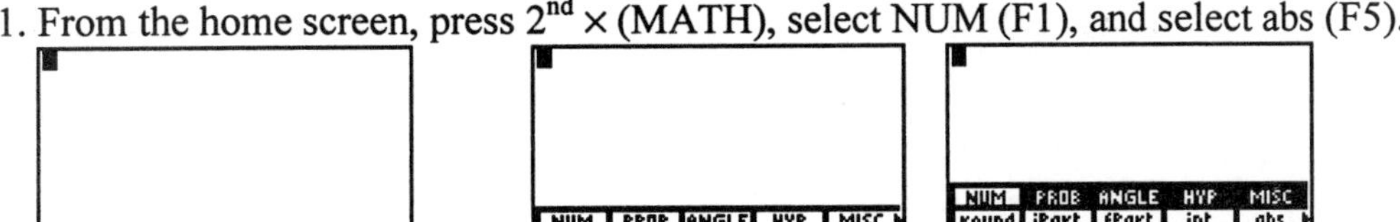

2. Enter –5 in the parentheses. Be sure to use the gray negative button not the black minus button.

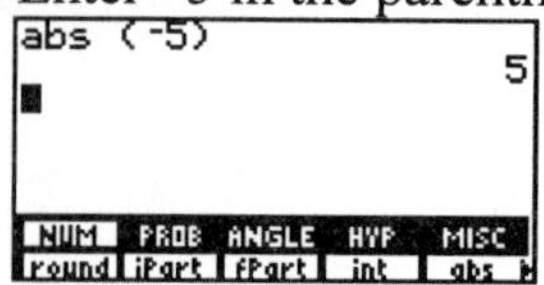

Example: Graph $y_1 = |x|$.

3. Press GRAPH and select y =(F1) and enter y_1 = abs(x). Follow the previous instructions for accessing the abs command.

4. Press ZOOM (M3) and select ZSTD (F4) to graph the function in the standard viewing window.

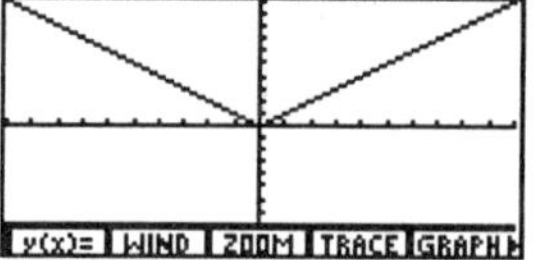

ANS

The ANS function recalls the previous answer so that additional operations can be performed on the answer to a previous operation.

Example: Start with the number 1 and continue to double it indefinitely.

1. Enter the number 1 and press ENTER. Press the multiplication button (×) then press 2 and ENTER. The ANS part of the display refers to the previous answer, in this case the 1 that was entered initially.

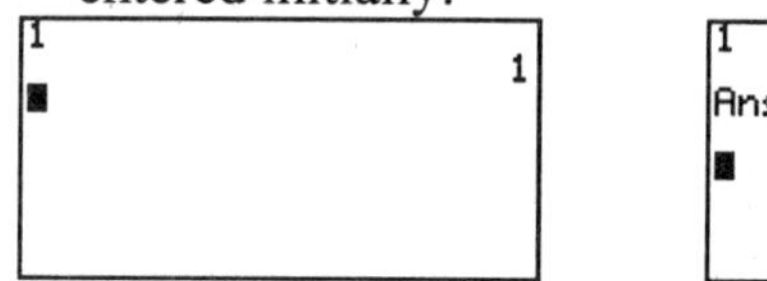

2. Continue to press ENTER until reaching the answer 64. Each time, the calculator takes the previous answer and multiples by 2.

The ANS function can be accessed by typing 2nd (-) (ANS).

3. Press 7, press +, press 2nd (-) (ANS), then ENTER. The calculator takes the last answer (64), adds 7, and returns the answer of 71.

Box and Whisker Plots

Box and whisker plots are used to display and compare data. Create a box and whisker plot for the data given. The ACT scores for some students in a particular school is given.

Student	1	2	3	4	5	6	7	8	9
ACT Score	20	22	30	32	28	21	20	34	18

1. Press 2nd + (STAT) and select EDIT (F2).

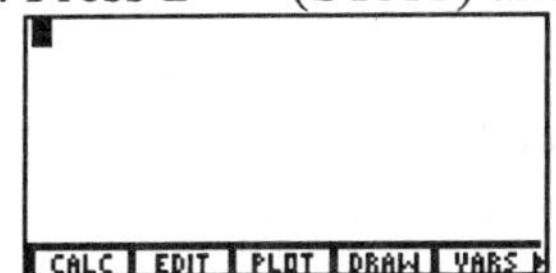

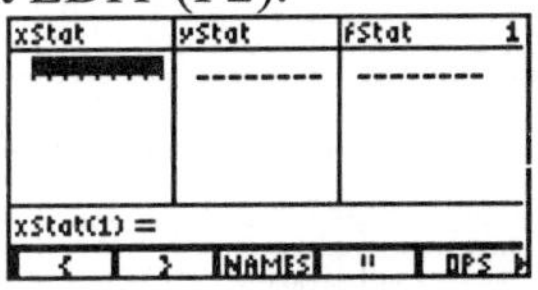

This will take you to the TI version of a spreadsheet called Lists. If data already is entered into the lists, you can clear it by using the UP ARROW to highlight the xStat at the very top of the screen. When highlighted, press CLEAR and ENTER. Repeat for yStat, etc. Enter the data into the lists by typing in the appropriate numbers and pressing ENTER after each entry. For this example, enter only the ACT scores into xStat.

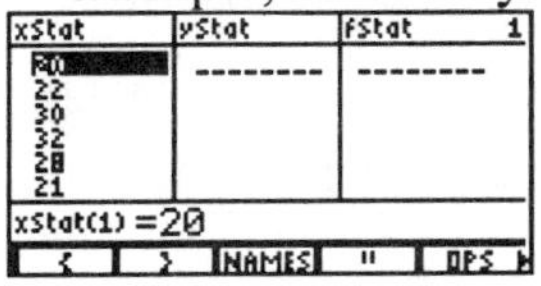

2. Set up the plot by pressing 2nd + (STAT). Choose PLOT (F3) and select Plot 1(F1), Plot 2 (F2) or Plot 3 (F3). Make sure that the plot is turned ON, the type is Box and Whisker (2nd to the last choice), the Xlist is the list containing the data (xStat in this case), and the frequency (Freq) is 1. Use the DOWN ARROW to move from one line to the next.

3. Adjust your WINDOW (press GRAPH and select WIND (F2)) so that it reflects the values entered into the lists. For a Box and Whisker plot, only the xMin and xMax must reflect the data. The yMin and yMax is irrelevant.

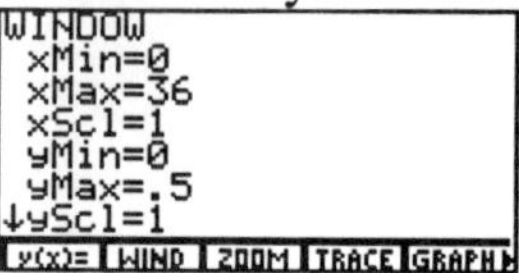

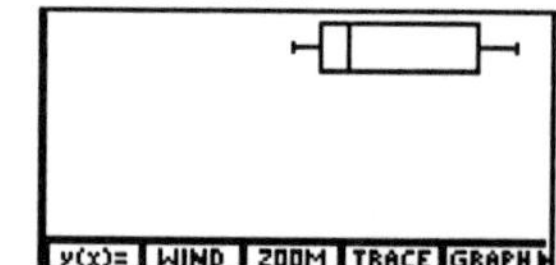

4. Use the TRACE (F4) command and your RIGHT and LEFT ARROW keys to find the minimum data value, first quartile, median, third quartile, and maximum data value.

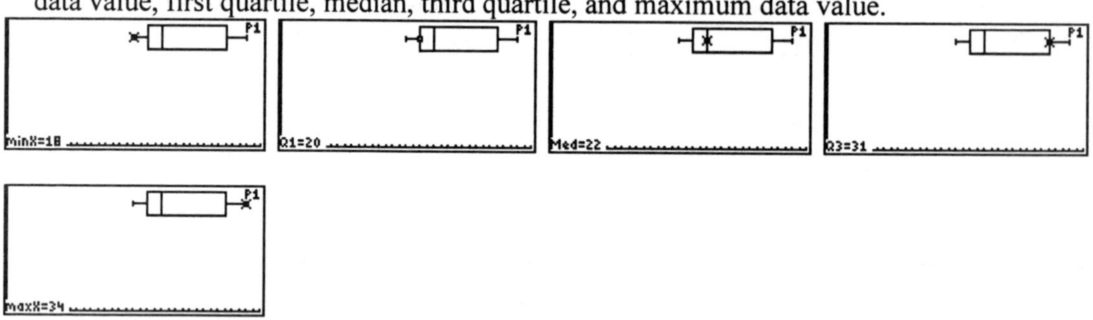

Contrast

The contrast for the calculator screen can be adjusted to be darker or lighter. As your batteries become older, it will become necessary to adjust the contrast so that you can more easily read the calculator screen.

1. Press 2nd and release.

2. Hold the DOWN ARROW key to make the screen lighter. Hold the UP ARROW key to make the screen darker. Each time the DOWN or UP ARROW key is released, the 2nd key must be pressed again to continue to adjust the contrast.

3. As the UP or DOWN ARROW keys are held, notice the number in the upper right hand corner. The contrast is measured from 1 – 9 where 1 is the lightest setting and 9 is the darkest. If the contrast needs to be 9 in order for you to see the screen, you probably need to replace the batteries.

DEL/INS

The DEL (Delete) key deletes the character under the current cursor location. The 2nd DEL (INS) (Insert) key inserts a character to the left of the current cursor location.

Example: You desire to type 9+6–5*34, but accidentally type 9+*36*–5**4*.

1. Enter *9+36–5*4*. Use the LEFT arrow to move the cursor to the 3 location.

9+36-5*4

2. Press DEL. This deletes the unneeded 3.

3. Use the RIGHT ARROW to move the cursor to the 4 location. Press 2nd DEL (INS) and enter a 3. This inserts the 3 to the left of the 4 as desired. Notice that the cursor changed after pressing 2nd DEL (INS) so that you knew that the calculator was in the insert mode.

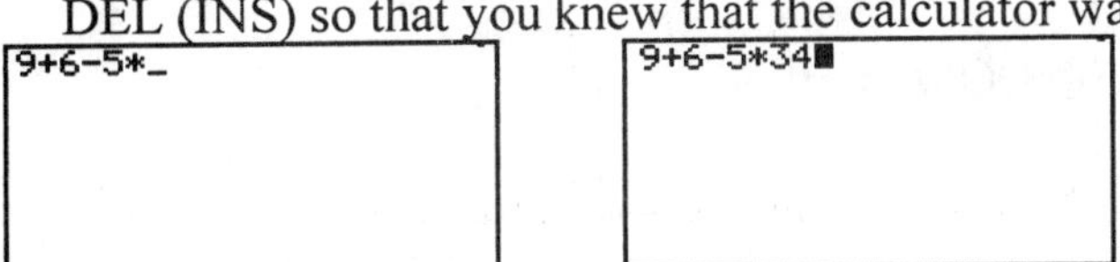

ENTRY

The ENTRY function returns previously entered operations on the calculator screen so that you may edit them.

1. Press 2nd ENTER (ENTRY). The last operation entered into the computer appears on the screen and the cursor is at the right end of the line. Use the UP, DOWN, RIGHT, LEFT arrow keys to edit the line.

2. Press 2nd ENTER (ENTRY) again. Continue to press 2nd ENTER (ENTRY) and notice that previous entries continue to be restored on the screen. Any one of them can be edited.

Error Messages

Occasionally, error messages are given to indicate some problem with the function you are asking the calculator to perform. Three common error messages and how to correct them are provided.

Syntax Error

Syntax refers to the way in which the function or command was entered. One common error deals with use of parentheses.

Example: You intend to enter 6*(3+2) but accidentally enter 6*3+2).

1. In the Home Screen, enter 6*3+2) and press ENTER.

6*3+2)■

ERROR 07 SYNTAX
GOTO QUIT

2. Because of the missing opening parenthesis, the calculator returns a syntax error. Notice the options at the bottom of the screen. You can QUIT and start over, or you can GOTO the error so that it can be corrected. Select GOTO (F1).

3. To correct the mistake, insert a parenthesis before the 3 using INS. Press ENTER.

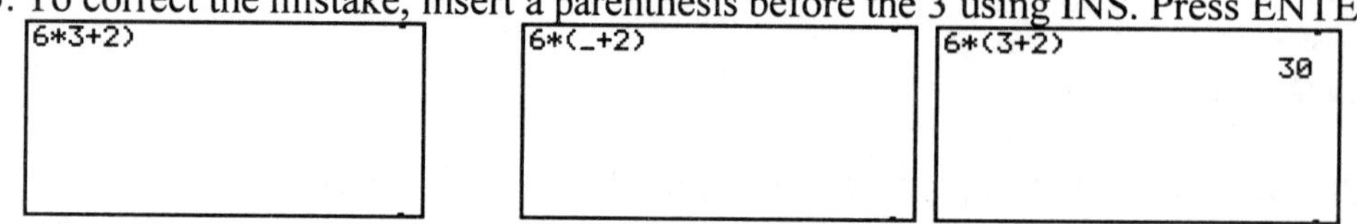

Graph Window Error

The Graph Window Error refers to an incorrect WINDOW entry.

1. Enter an equation into $y =$.
   For example: $y_1 = x^3 + 3$

2. Press ZOOM (M3) and select ZSTD (F4) to graph the function in the standard viewing window.

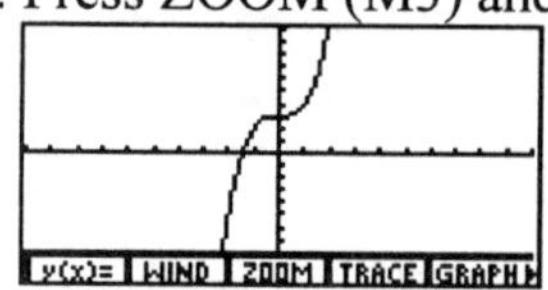

3. Press WIND (F2). Imagine that you incorrectly changed the WINDOW as shown below. Press GRAPH (F5). A Graph Window Error occurs.

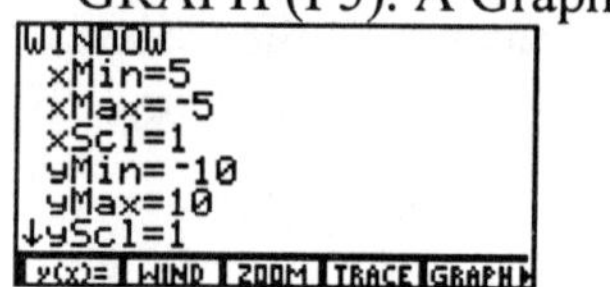

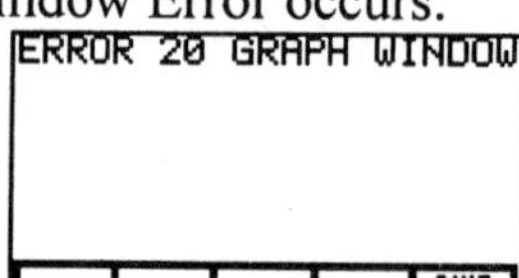

4. The only option is to QUIT (F5). To correct the error, press GRAPH, select WIND (F2) and change the xMin and xMax as shown. Press GRAPH (F5).

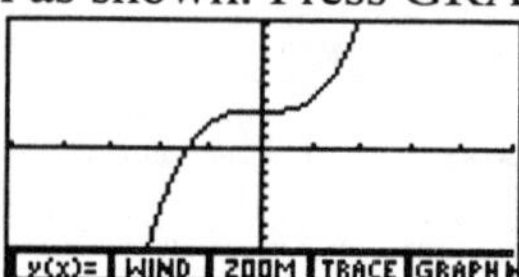

Dimension Mismatch Error (DIM MISMATCH)

The Dimension Mismatch error results when the data entered in xstat and ystat do not match up. That is, there are more data entered in one list than in the other.

1. When DIM MISMATCH is encountered, the only option is to QUIT (F5).

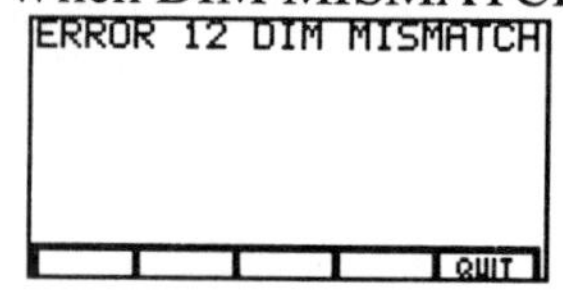

2. Press QUIT (F5). To correct the error, refer to Preparing the Calculator, step 2.

Evaluate a Function

One way to evaluate a function is to use the notation $y_1(x)$ in the home screen.

1. Enter an equation into y_1.
   Example: $y_1 = 2x^2 + 3x + 1$

2. Press 2nd EXIT (QUIT) to go back to the home screen. Press 2nd CUSTOM (CATLG-VARS). Press MORE. Press EQU (F4). Use the arrow keys to move the cursor to the desired equation, in this case y_1, and press ENTER.

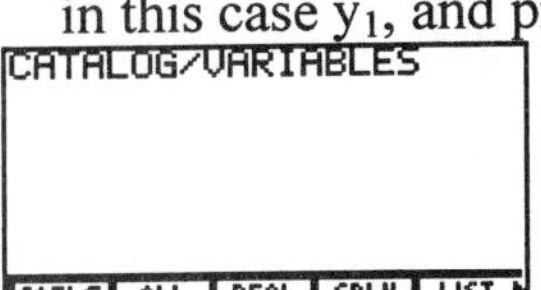

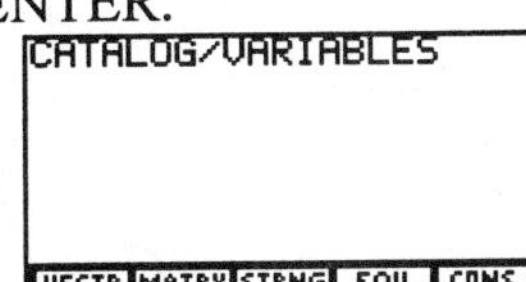

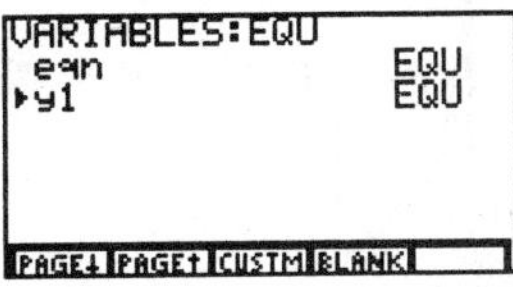

3. Type a left parenthesis, enter a value for x, close the parenthesis and press ENTER.

EXIT

The EXIT key will back out of a menu one screen at a time.

Example: Press GRAPH then ZOOM (F3). To get back to the home screen, press EXIT to return to the GRAPH menu, then EXIT again to return to the home screen.

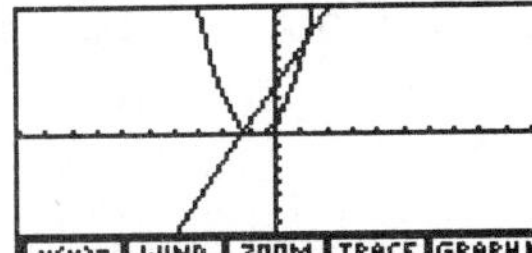

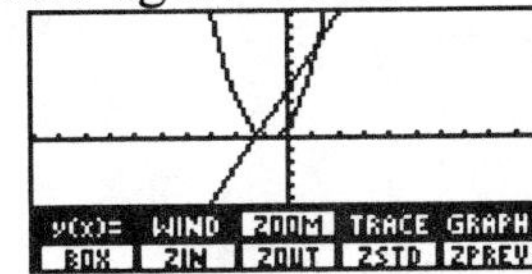

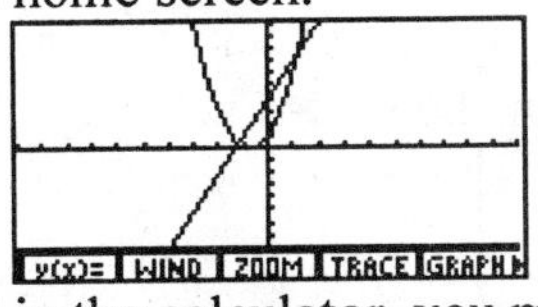

Note: Depending upon what equations, if any, are entered in the calculator, you may see a different graph than shown above.

Exponents

Exponents of 2 can be entered into the calculator 2 different ways. Exponents of greater than 2 are entered using the caret key.

Example: Evaluate $3^2 + 7^2 - 8^4 + 2^7$.

1. To enter the powers of 2, use the x^2 key that is located in the middle of the first column of keys. To enter powers greater than 2, use the caret key (∧) which is located below the CLEAR key in the last column.

```
3²+7²-8^4+2^7
```

2. Press ENTER.

```
3²+7²-8^4+2^7
                 -3910
```

Fractions

The TI-86 can convert from fractions to decimals and from decimals to fractions.

Example: Add $\frac{3}{8}+\frac{5}{12}$. Write your answer as a fraction.

1. Enter 3/8 + 5/12 and press ENTER. The fraction bar (/) is the division symbol on the keypad.

```
3/8+5/12
         .791666666667
```

2. To change the answer from a decimal to a fraction, press 2nd × (MATH) and select MISC (F5). Press MORE, then choose ▶Frac (F1).

```
3/8+5/12
         .791666666667
Ans▶Frac
                 19/24
NUM  PROB  ANGLE  HYP  MISC
▶Frac  %  pEval  ×√  eval
```

Note: See the ANS section if you don't know what Ans means.

3. To convert the fraction back to a decimal, you must type in 19/24 and press enter.

```
         .791666666667
Ans▶Frac
                 19/24
19/24
         .791666666667
NUM  PROB  ANGLE  HYP  MISC
▶Frac  %  pEval  ×√  eval
```

This screen shows that the fraction $\frac{19}{24}$ is equivalent to the decimal 0.7916666667.

GRAPH and $y =$

The GRAPH and $y =$ menu is used to enter functions into the calculator that will be graphed.

1. Press GRAPH and $y =$ (F1).

2. Up to 99 equations can be entered and graphed. After entering an equation, press ENTER to enter another equation.

3. Type in the equation to be graphed.
   Example: $y_1 = -4x + 5$ (Note: Be sure to use the gray (-) key not the black – key when typing the $-4x$ part of the equation.)

4. Press 2nd GRAPH (M5). Depending on the viewing window, you may or may not see the graph. To view the graph in the standard viewing window, press ZOOM (F3) and select ZSTD (F4). See the ZOOM section of the manual for more details.

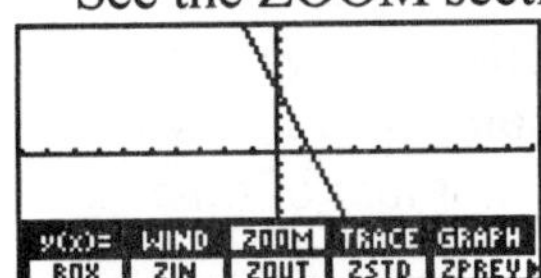

5. Press CLEAR to remove the menu choices from the bottom of the screen.

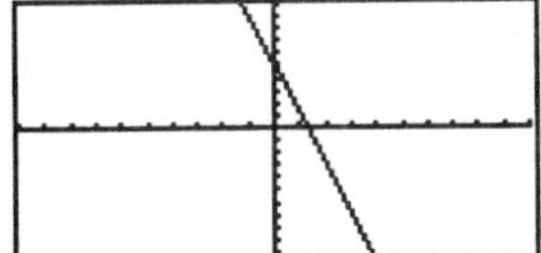

Histograms

Histograms are a way to display and analyze data. Create a histogram for the data given.

The shoe size of some the children in the class is shown.

Child	1	2	3	4	5	6	7	8	9
Shoe Size	5	5	4	5	4	3	6	5	4

1. Press 2nd + (STAT) and select EDIT (F2).

This will take you to the TI version of a spreadsheet called Lists. If data already is entered into the lists, you can clear it by using the UP ARROW to highlight the xStat at the very top of the screen. When highlighted, press CLEAR and ENTER. Repeat for yStat, etc. Enter the data into the lists by typing in the appropriate numbers and pressing ENTER after each entry. For this example, enter only the shoe size into xStat.

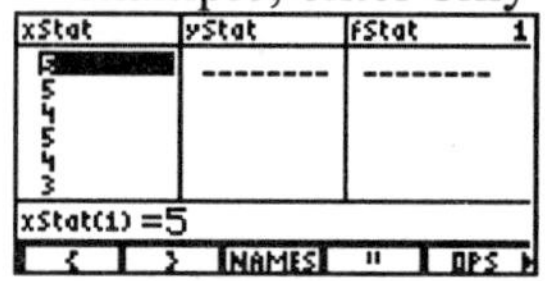

2. Set up the plot by pressing 2nd + (STAT). Choose PLOT (F3) and select Plot 1(F1), Plot 2 (F2) or Plot 3 (F3). Make sure that the plot is turned ON, the type is Histogram (HISTO) (2nd to the last choice), the Xlist is the list containing the data (L_1 in this case), and the frequency (Freq) is 1. Use the DOWN ARROW to move from one line to the next.

3. Adjust your WINDOW (press GRAPH and select WIND (F2)) so that it reflects the values entered into the lists. The yMin should be set at 0 because 0 is the least number of times a particular shoe size is seen. The yMax should be set at 5 since the largest number of times a particular shoe size is seen is 4 (4 children wear size 5 shoes).

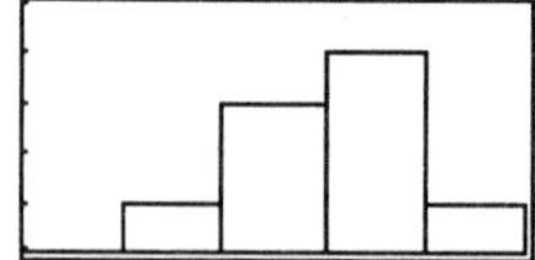

4. Use the TRACE (press GRAPH and select TRACE (F4)) command and your RIGHT and LEFT ARROW keys to find the frequency of each shoe size.

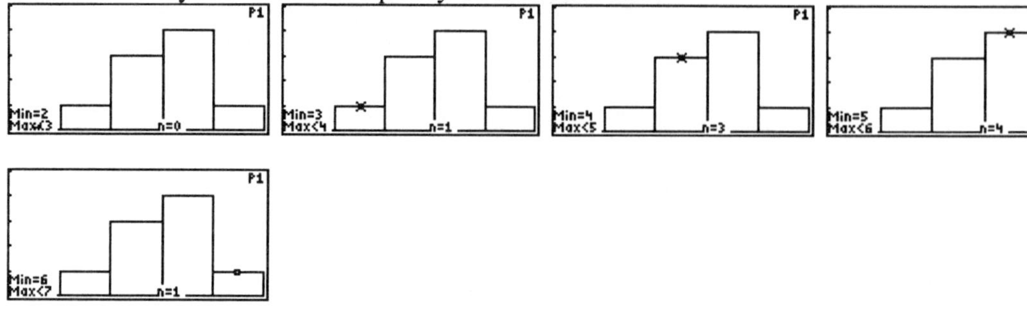

ISECT

The ISECT (intersect) function will calculate an intersection point of two graphed equations.

1. Enter the equations in y_1 and y_2.

   Example: $y_1 = 2x^2 + 3x + 1$
   $y_2 = 3x + 4$

2. Select an appropriate viewing window. In this case the standard viewing window may be used (ZSTD).

3. Either press GRAPH (M5) or ZOOM (M3) and select ZSTD (F4).

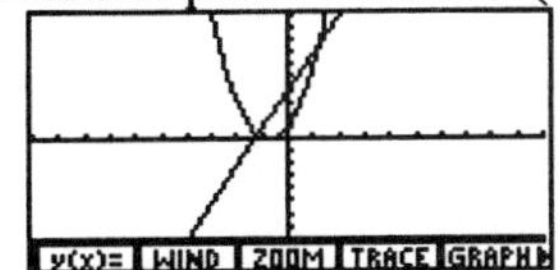

4. To determine the first quadrant point of intersection for these two graphs, press MORE, MATH (F1), MORE, ISECT (F3).

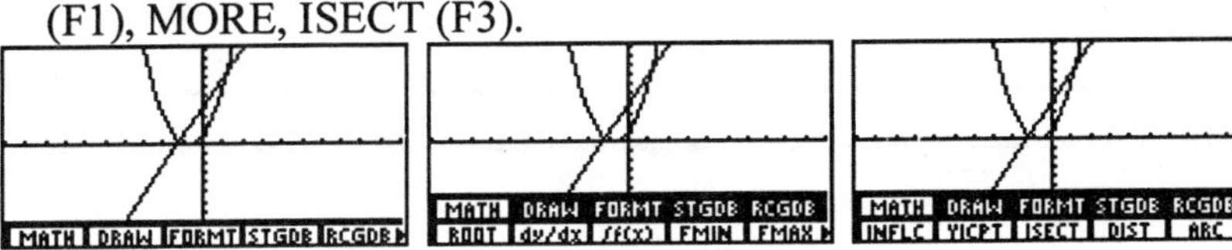

5. Automatically returning to the graph screen, the calculator asks for the First curve. The First curve is the function defined in Y_1. At the top of the screen, a 1 should be showing, designating the fact that the first curve is in Y_1. If it is, press ENTER. If not, press either the UP or DOWN arrow to switch to that equation, then press ENTER.

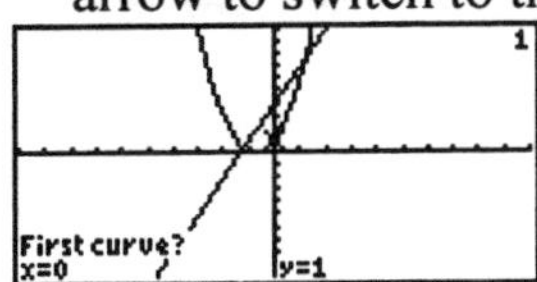

6. The calculator now asks for the Second curve. The Second curve is the function defined in Y_2. At the top of the screen, a 2 should be showing. If it is, press ENTER. If not, press either the UP or DOWN arrow to switch to that equation, then press ENTER.

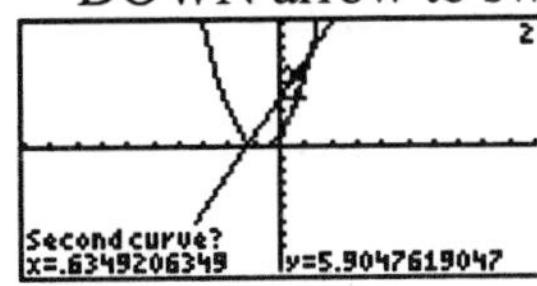

7. The calculator next asks for a Guess of the point of intersection. Use the LEFT and RIGHT arrow keys to move the cursor to the place where the graphs intersect in the first quadrant (in this example). Press ENTER.

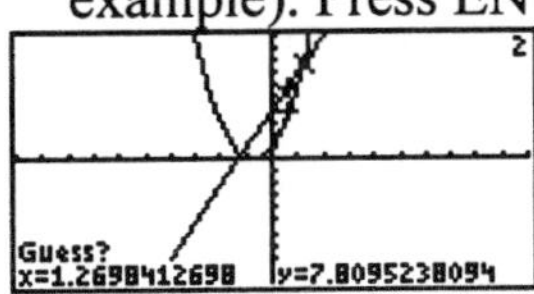

8. The calculator will show the point of intersection at the bottom of the screen.

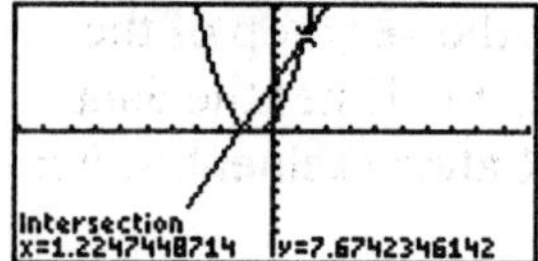

Lists – See Regression

ON/OFF

1. Press ON to turn the calculator on.

2. To turn the calculator off, press 2nd ON (OFF).

After several minutes of inactivity, the calculator will automatically shut off.

One-Variable Statistics

The 1-Variable Statistics command returns the following statistics: mean ($\bar{x}$), sum of x ($\sum x$), sum of the squares of x ($\sum x^2$), population standard deviation (σx), the amount of data entered (n), the minimum data value (minX), the first quartile (Q1), the median (Med), the third quartile (Q3), and the maximum data value (maxX).

Example: Determine all of the one-variable statistics for the following data set.

The ACT scores for some students in a particular school is given.

Student	1	2	3	4	5	6	7	8	9
ACT Score	20	22	30	32	28	21	20	34	18

1. Press 2nd + (STAT) and select EDIT (F2).

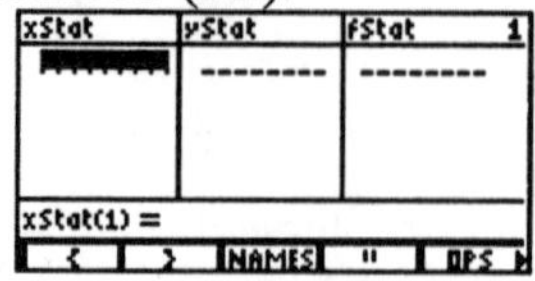

This will take you to the TI version of a spreadsheet called Lists. If data already is entered into the lists, you can clear it by using the UP ARROW to highlight the xStat at the very top of the screen. When highlighted, press CLEAR and ENTER. Repeat for yStat, etc. Enter the data into the lists by typing in the appropriate numbers and pressing ENTER after each entry. For this example, enter only the ACT score into xStat.

2. From the home screen (press 2nd EXIT (QUIT)), press 2nd + (STAT) and select CALC (F1). Select OneVa (F1). You must tell the calculator which list you want to use to compute the statistics. Press 2nd – (LIST), choose NAMES (F3), and choose the list containing the data (xStat (F2) in this case). Press ENTER.

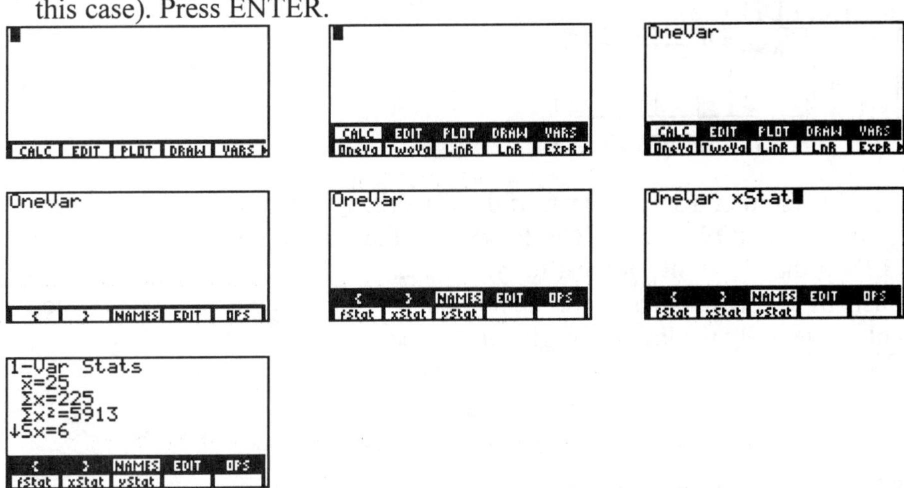

3. Notice the arrow at the bottom of the screen. This indicates that there is more information to be seen. Use the DOWN ARROW to view this information.

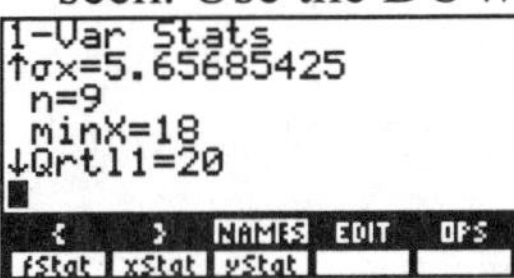

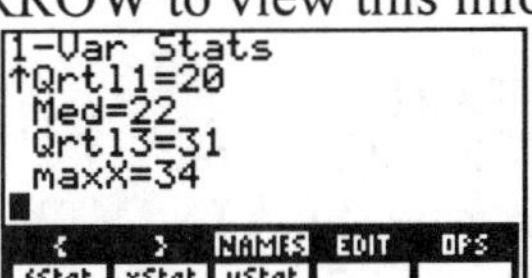

QUIT

This returns the calculator to the home screen from any other menu.

Example: Imagine that you are in the WINDOW screen and need to get back to the home screen.

1. Press WINDOW. To go back to the home screen, press 2nd EXIT (QUIT).

Regression

The regression feature creates a mathematical model for data that has been entered in the Lists. For example, create a scatter plot and a mathematical model for the set of data given.

The revenue for Dell Computer Corporation from 1985 through 1993 is listed in the table. The revenue is given in millions of dollars.

Year	1985	1986	1987	1988	1989	1990	1991	1992	1993
Revenue	33.7	69.5	159.0	257.8	388.6	546.2	889.0	2013.9	2873.2

(Source: Dell Computer Corporation)

1. Push 2nd + (STAT) and select EDIT (F2).

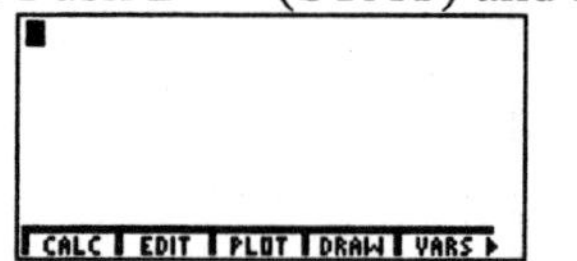

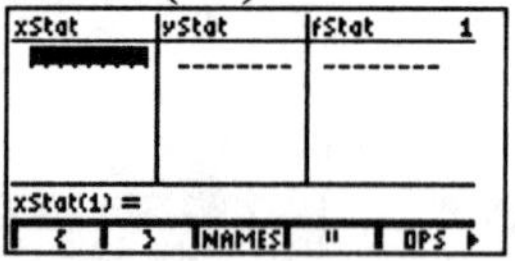

This will take you to the TI version of a spreadsheet. If data is already entered into the lists, you can clear it by using the UP ARROW to highlight the xStat at the very top of the screen. When highlighted, press CLEAR then ENTER. Repeat for yStat, etc.

Enter the data into the lists by typing in the appropriate numbers and pressing ENTER after each entry. For this example, enter only the last two digits of the year.

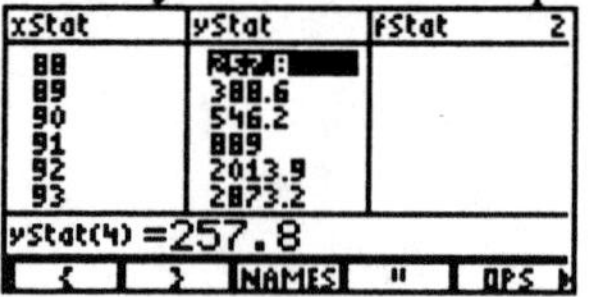

2. Set up the plot by pressing 2nd STAT again.

Select PLOT (F3). Select Plot 1, 2 or 3 (F1, F2 or F3).

Press ENTER when the cursor is on the word ON to turn the plot ON. Use the DOWN arrow to move to the Type option. Select scatter plot for type (SCAT). DOWN arrow to Xlist and choose the Xlist and Ylist names (usually xStat and yStat), and select the MARK that you prefer. (Note: the options for each of these appears at the bottom of the screen as you arrow down to each option.)

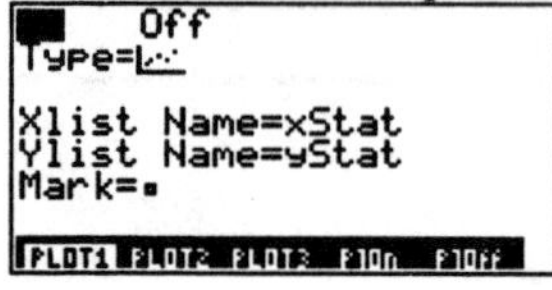

3. Adjust your WINDOW by pressing GRAPH and WIND (F2) so that it includes the values entered into your lists.

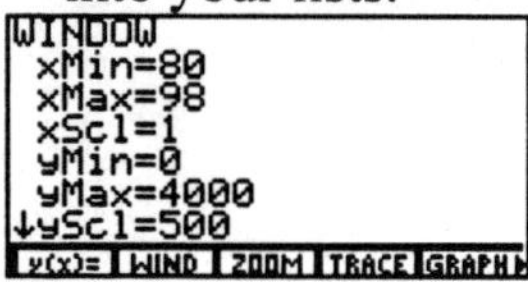

4. Press GRAPH (F5). To see the data points hidden behind the menu area, press CLEAR.

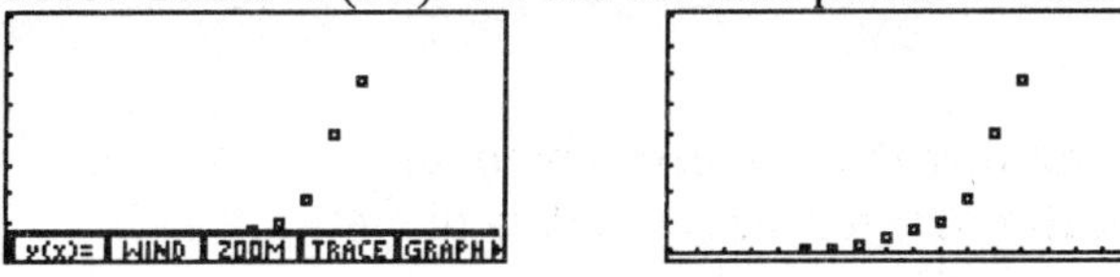

5. To have the calculator determine an equation of a curve that models the data, press 2nd STAT again and choose CALC (F1). At the bottom of the screen are many options for regression equations.

For example: Select ExpR (F5) (Exponential Regression). This puts the command into the home screen. Tell the calculator what lists you want it to work with by pressing 2nd LIST (above the – sign), choosing NAMES (F3), and the name of the list in which your x data is located (usually xStat). Press the comma button (to the left of the number 4) then the name of the list where your y data is located (usually yStat).

6. Press enter and the calculator will return the equation of the line in the form $y = a \cdot b^x$.

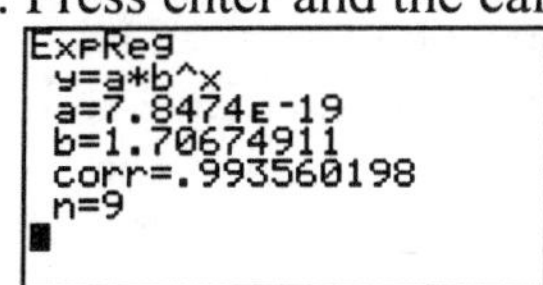

You can record this information on your paper and enter the equation in y =, or…

7. To graph the function without needing to remember or write down the regression model, press GRAPH and select y = (F1). Press 2nd CUSTOM (CATLG-VARS), press MORE, MORE, and select STAT (F4).

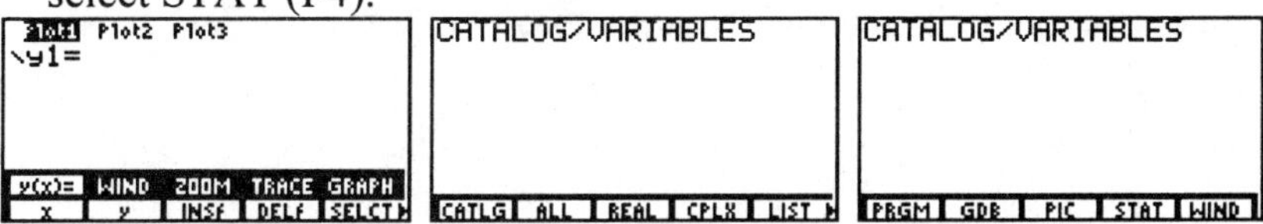

8. Use the DOWN arrow to select the RegEq option. Press ENTER. This pastes the RegEq command into the y = menu.

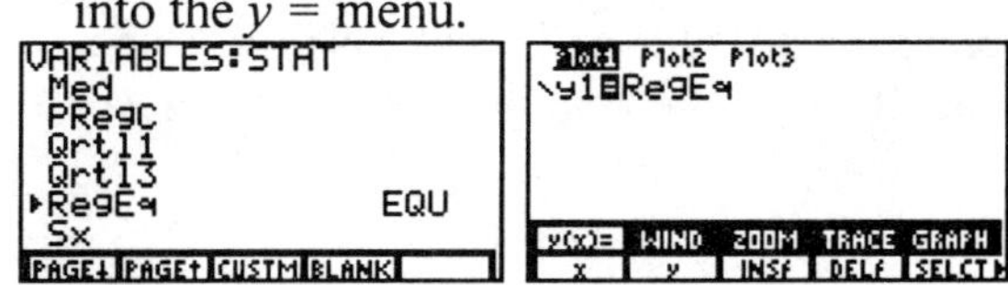

9. Press GRAPH (M5). Press CLEAR to view the entire graph.

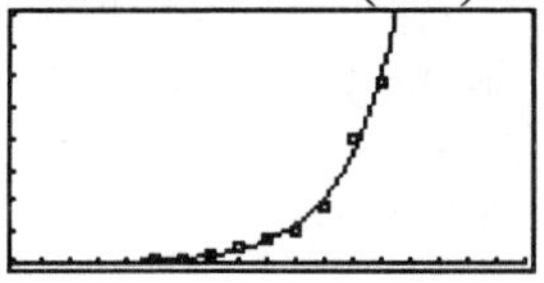

Root

The Root function calculates the x-intercepts of a function in the graphing screen.

1. Enter a function in the GRAPH, $y =$ menu.
   For example: $y_1 = x^2 + 3x - 5$

2. Press 2nd ZOOM (M3) and select ZSTD (F4) to view the function in the standard viewing window. Press CLEAR to remove the menu choices from the bottom of the screen.

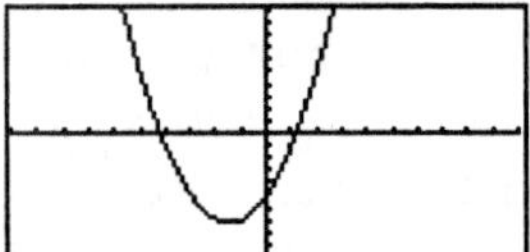

3. Press GRAPH then MORE and select MATH (F1). Select ROOT (F1).

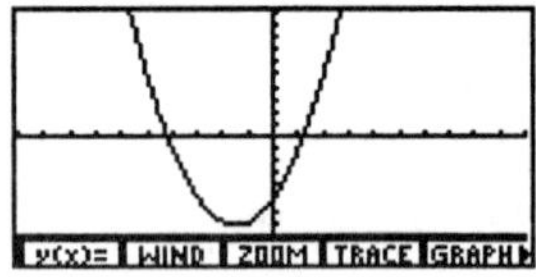

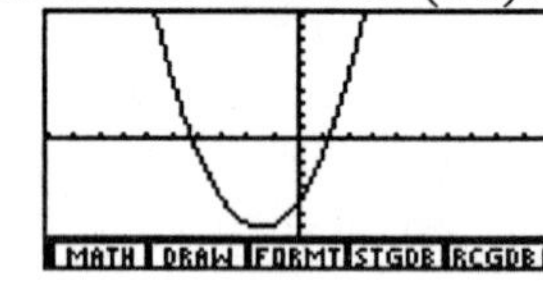

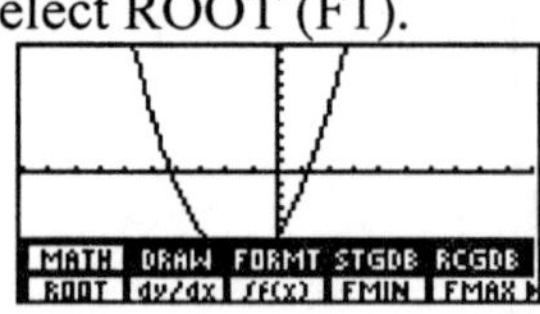

4. The root function calculates one x-intercept at a time. After selecting root, the calculator asks for the Left Bound. This is a value to the left of the x-intercept being calculated. Use the LEFT and RIGHT arrow keys to move the cursor somewhere to the left of the zero being calculated and press ENTER.

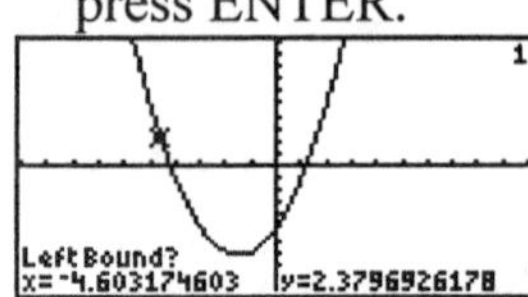

5. The calculator asks for the Right Bound. Use the LEFT and RIGHT arrow keys to move the cursor somewhere to the right of the root being calculated and press ENTER.

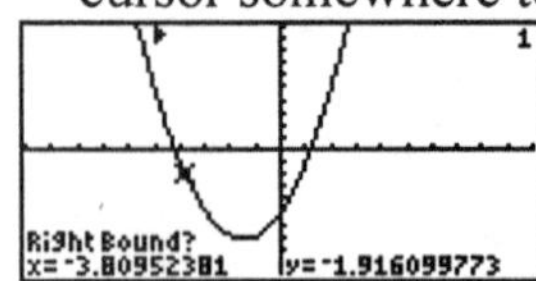

6. The calculator asks for a guess as to where the x-intercept is located. Use the LEFT and RIGHT arrow keys to move the cursor as close to the x-intercept as possible and press ENTER. The x-intercept should now be displayed. Notice the y-value. It should be zero or at least very close to zero!

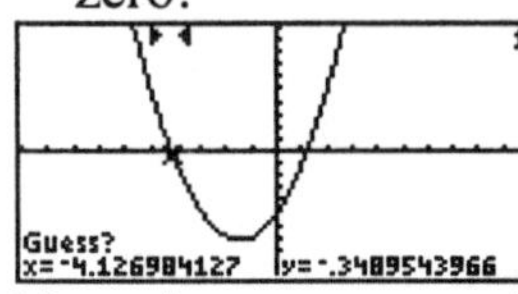

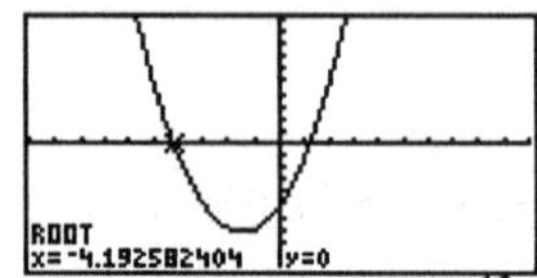

   Note: In some cases the y-value could be -1×10^{-12}, for example, which is extremely close to zero!

7. To find the other root, repeat steps 3-6 focusing on the x-intercept on the positive x-axis.

Scatter Plots – See Regression

Scientific Notation

The calculator will automatically go in to scientific notation when the result of an operation is either very small or very large.

Example: It is estimated that a satellite orbiting the earth has traveled 10,000,000 miles. How many centimeters is that?

1. Enter 10,000,000*5280*12*2.54 then press ENTER. (Note: There are 5280 feet in one mile, 12 inches in one foot, and 2.54 centimeters in one inch.)

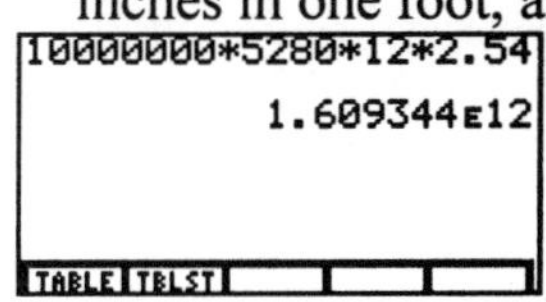

2. Notice the notation used by the calculator. The 1.609344E12 means 1.609344×10^{12}, or 1,609,344,000,000 centimeters.

Square Root

The calculator will calculate the square root of a number and return the positive root, accurate up to 11 decimal places.

Example: What is the square root of 111?

1. Enter the square root function first. Press 2nd x^2 ($\sqrt{\ }$). Type then number that you which to calculate the square root, in this 111. Press ENTER.

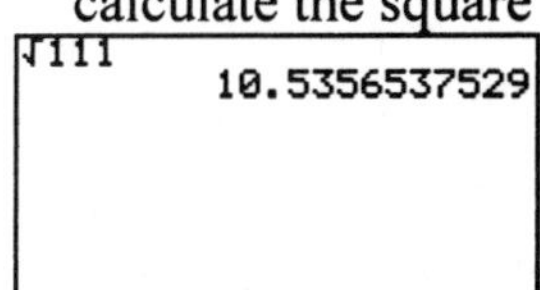

Note: Parentheses are needed if you need to calculate the square root of the sum or difference of numbers.

Example: What is the square root of 23 + 79?

2. Press 2nd x^2 ($\sqrt{\ }$). Press the left parenthesis. Enter 23+79 and close the parentheses. Press ENTER.

√(23+79)
10.0995049384

Tables

The table feature allows the user to view a table of values when the equation of the function is entered in the $Y =$ menu.

1. Enter an equation into y_1.

   Example: $y_1 = 2x^2 + 3x + 1$

2. Press TABLE. Select TBLST (F2).

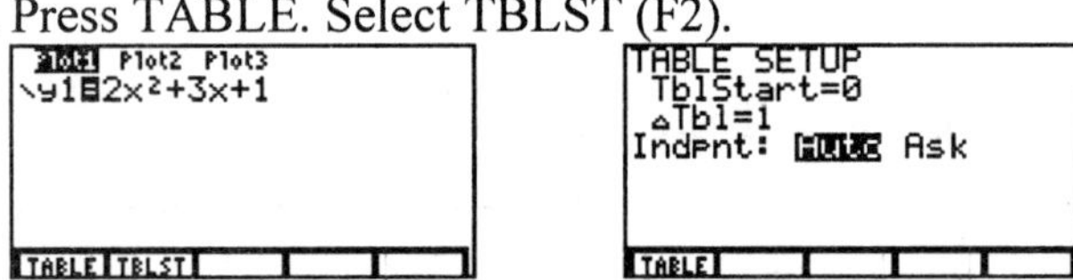

3. TblStart determines where the table values will begin.

   Example: TblStart=0

4. ΔTbl determines the interval between table entries. For example, if increments of one are desired, enter a 1 here.

   Example: ΔTbl=1
   Note: Δ is the Greek letter delta, which stands for change. So this is literally the change in the table.

5. To view a complete table of values which can be scrolled through up or down, leave the Indpnt option on Auto.

6. To view the table, press TABLE (F1).

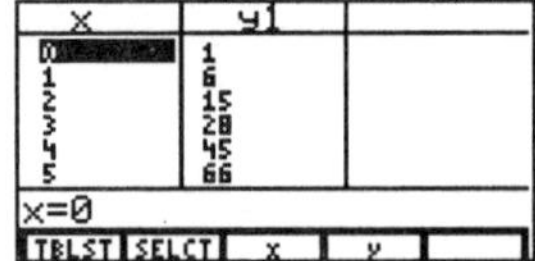

7. You can scroll up or down through the table by pressing and holding the UP or DOWN arrow keys. You must have the cursor in the x column in order to use the scrolling feature.

   NOTE: The values in the y1 column of the table may be rounded. To get better accuracy, arrow over to the value in the y1 column and look at the bottom of the screen.

Another way to evaluate a function is to change the Indpnt option to Ask.

8. Repeat steps 2-4. This time, change the Indpnt option to Ask by using the UP, DOWN, LEFT, and RIGHT ARROWS to move the cursor to the Ask option and pressing ENTER.

9. Press TABLE (F1). Enter values into the X column and the calculator will display the corresponding y-value.

Example: Find the value of the function when $x = 5$.

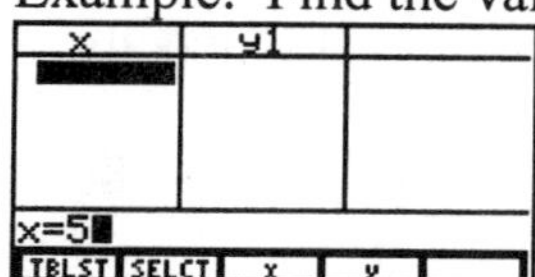

10. You can continue to evaluate the function in Y_1 by continuing to enter values for x.

TRACE

The TRACE feature allows you to trace along the graph of a function and view the coordinates of the pixels at the bottom of the screen.

Example: Graph the function $y = x^3 + 2x^2 - 3x - 4$ and use the TRACE feature to approximate the x-intercepts of the graph.

1. Enter the equation into y_1 using the $y =$ screen (press GRAPH and select $y =$ (F1)). Press ZOOM (M3) and select ZSTD (F4) to graph the function in the standard viewing window.

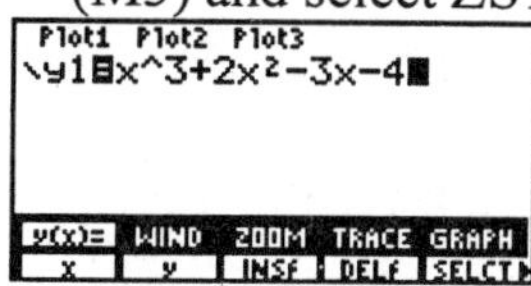

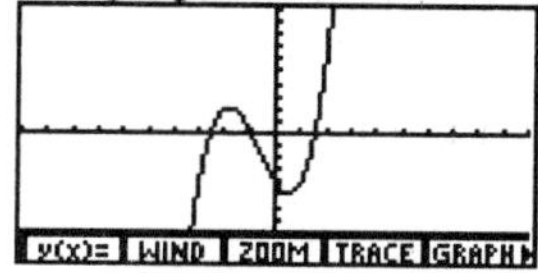

2. Press TRACE (F4). Use the LEFT and RIGHT ARROW keys to move the cursor along the graph of the curve. Estimate the zeros of the function by moving the cursor as close to the x-intercepts as possible.

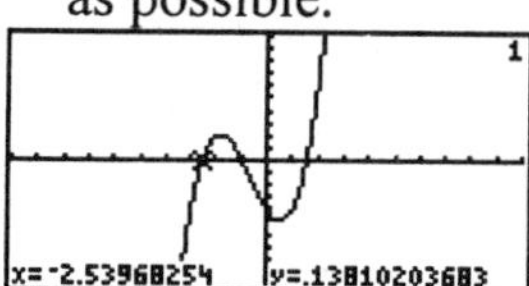

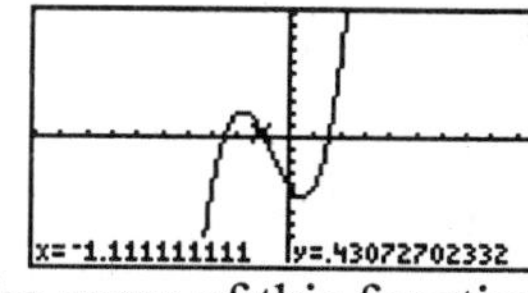

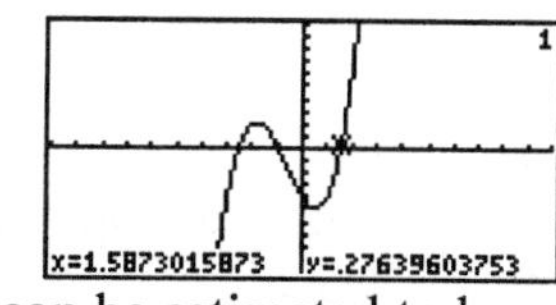

Using the TRACE feature, the zeros of this function can be estimated to be $x = -2.6, -1.1$, and 1.6.

Trigonometric Functions

The TI-86 is capable of calculating the sine, cosine, and tangent ratios. The inverse trigonometric functions can also be calculated. For this section, be sure that your calculator is in degree mode.

1. Press 2nd MORE (MODE). If necessary, use the DOWN ARROW to move to the Radian Degree line and highlight Degree by moving the cursor to Degree and pressing ENTER.

Example: Use trigonometric and inverse trigonometric functions to solve the right triangle.

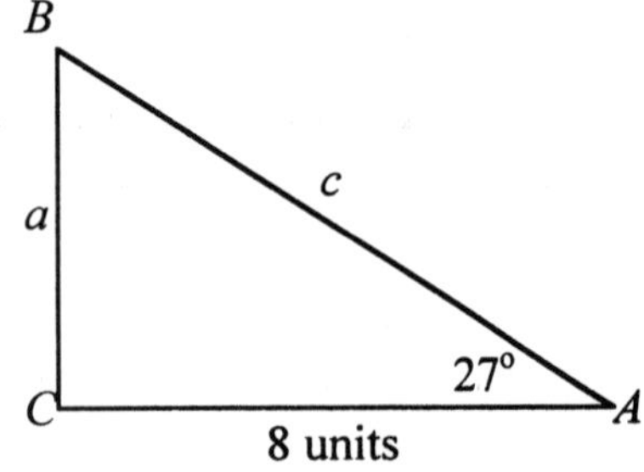

1. To determine the measure of side c, use the cosine ratio.

$$\cos 27° = \frac{8}{c}$$

$$c = \frac{8}{\cos 27°}$$

2. Press 8 ÷ COS 27 ENTER.

```
8/cos (27)
        8.97860990107
```

Note: The left parenthesis is not automatically inserted on the TI-86.

3. To determine the length of side a, use the tangent ratio.

$$\tan 27° = \frac{a}{8}$$

4. Press 8*tan 27 ENTER.

```
8*tan (27)
        4.07620359596
```

5. To determine the measure of angle B, use the inverse sine function.

$$\sin B = \frac{8}{c}$$

$$\sin B = \frac{8}{8.97}$$

$$B = \sin^{-1}\frac{8}{8.97}$$

6. Press 2nd SIN (SIN^{-1})(8 ÷ 8.97) ENTER.
NOTE: The ratio is entered in parenthesis.

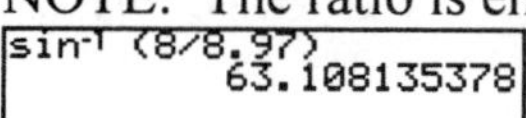

Value

The value feature can be used to evaluate a function at a particular value of x. It works in conjunction with a graph of a function.

1. Enter an equation into y_1.
Example: $y_1 = 2x^2 + 3x + 1$

2. Press ZOOM (M3) and ZSTD (F4) to view the function in the standard viewing window.

3. To evaluate the function at different values of x, press MORE twice and select EVAL (F1).

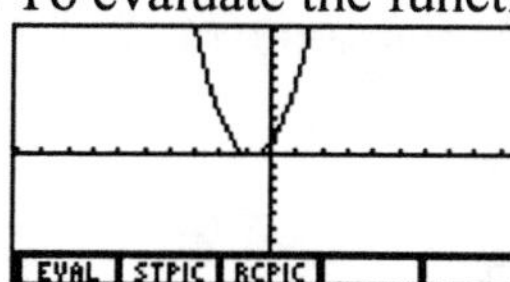

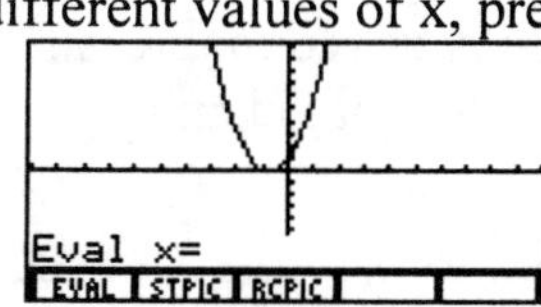

4. Enter a value for x for which you would like to find the value of the function.
Example: x=2

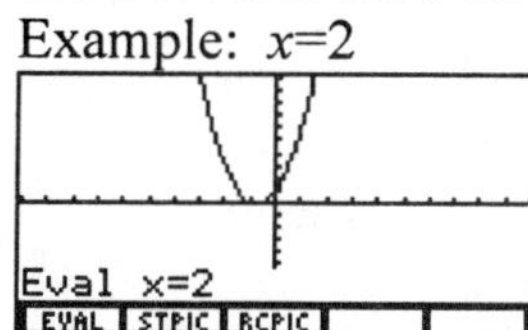

5. Press ENTER to see the value of the function at the selected value of x.

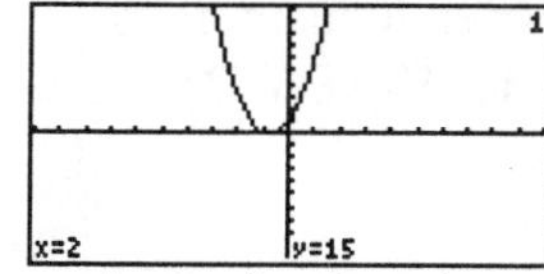

To evaluate the function at another value of x, press GRAPH and repeat steps 3-5.
NOTE: This function only works for x values in the WINDOW.

ZOOM/WINDOW

The ZOOM button changes the viewing window rapidly with pre-set ZOOM functions.

1. Enter an equation into the y = screen.

   Example: $y_1 = 2x^2 + 3x + 1$.

2. Press 2nd ZOOM (M3) and select ZSTD (F4). ZSTD creates the standard viewing window which goes from –10 to 10 on the x-axis and –10 to 10 on the y-axis. Press WIND (F2) to see this.

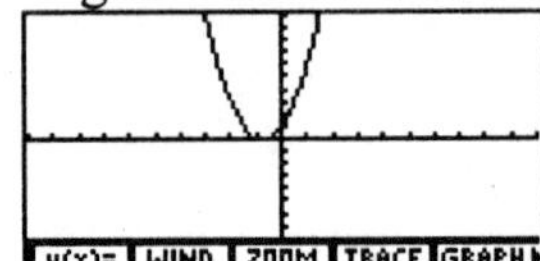

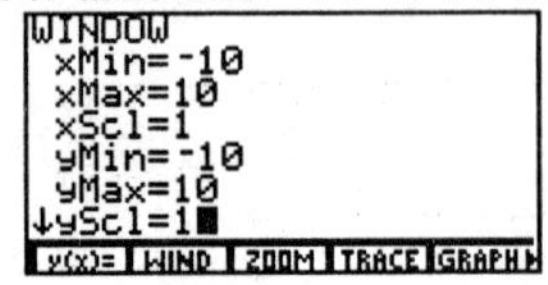

3. Press ZOOM (F3) and select BOX (F1). BOX allows zooming in on a particular region on the graph by boxing in the desired region. After pressing BOX, use the UP, DOWN, RIGHT, and/or LEFT arrow keys to move the cross hair around the screen. Think about where you would like one corner of the boxed region to be. When you have moved the cross hair to that spot, press ENTER.

   For Example:

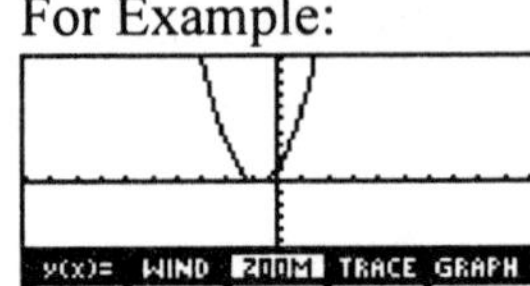

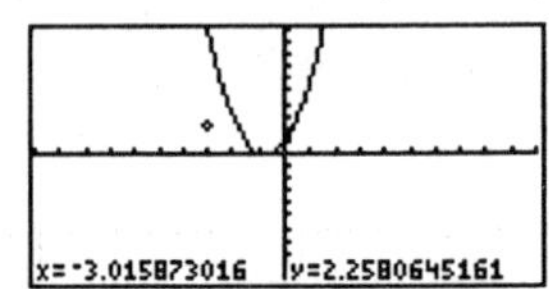

4. Use the UP, DOWN, RIGHT, and/or LEFT arrow keys to create a box around the region in which you wish to ZOOM. When the desired region is boxed in, press ENTER. When completed, press WIND (F2) to see how the viewing window has changed.

   For Example:

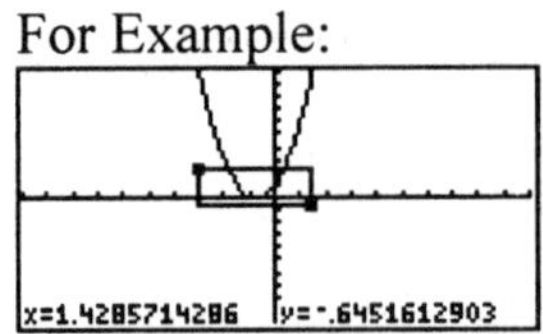

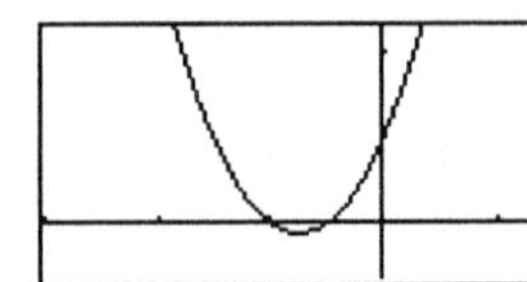

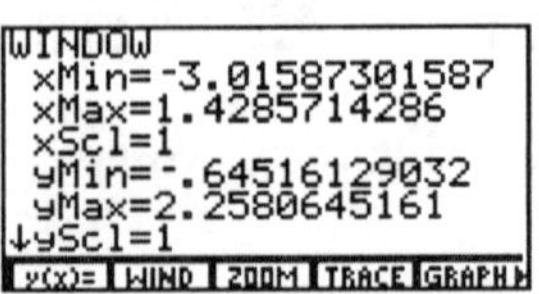

5. Press ZOOM (F3) and select ZSTD (F4) to return the graph to the standard viewing window.

6. Press ZOOM (F3) and select ZIN (F2). ZIN will do zoom in, centered at a point that you select. After pressing ZIN (F2), use the UP, DOWN, RIGHT, and/or LEFT arrow keys to move the cross hair around the screen. Stop when the desired center of the ZIN has been located. Press ENTER to cause the Zoom In to occur. When completed, press GRAPH and WIND (F2) to see how the viewing window has changed.

   For Example:

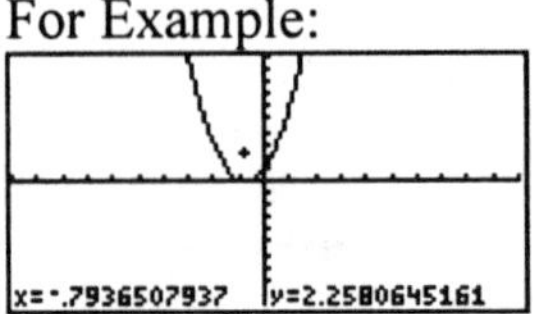

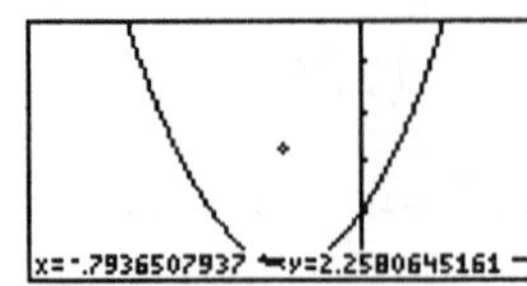

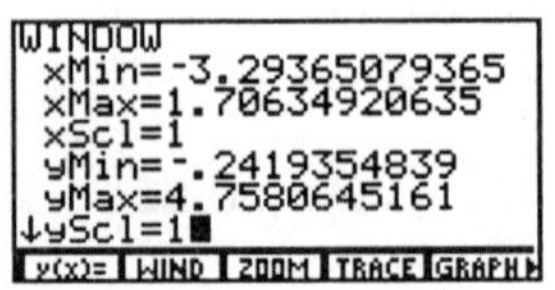

Note: Your WINDOW values may be different than this example.

7. Press ZOOM (F3) and select ZSTD (F4) to return the graph to the standard viewing window.

8. Press ZOOM (F3) and select ZOUT (F3). ZOUT will zoom out, centered at a point that you select. After pressing ZOUT (F3), use the UP, DOWN, RIGHT, and/or LEFT arrow keys to move the cross hair around the screen. Stop when the desired center of the ZOUT has been located. Press ENTER to cause the Zoom Out to occur. When completed, press GRAPH and WIND (F2) to see how the viewing window has changed.

For Example:

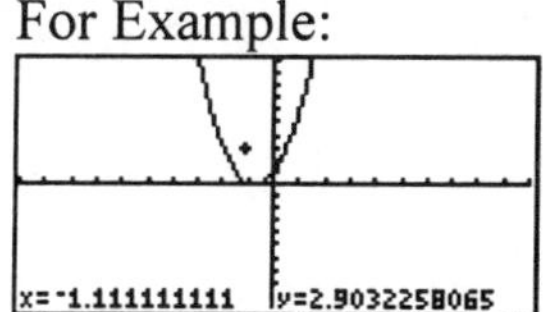

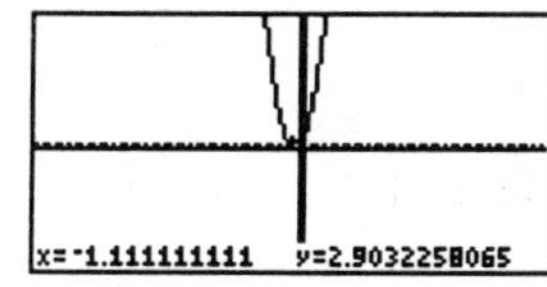

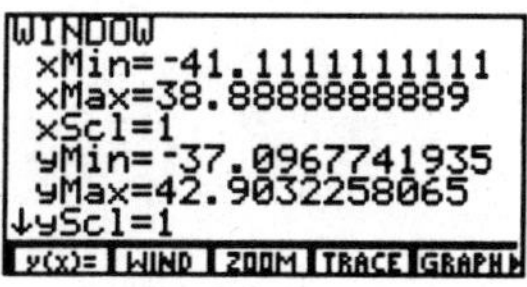

9. Press ZOOM (F3) and select ZSTD (F4) to return the graph to the standard viewing window.

10. Press ZOOM (F3) then MORE and select ZDECM (F4). ZDECM will automatically create a viewing window which is considered "friendly". This means that when you TRACE on a graph viewed using ZDECM, the values shown will be "friendly" decimals, showing accuracy to the nearest tenth rather than the long decimal values shown in other viewing windows. When completed, press WIND (F2) to see how the viewing window has changed.

For Example:

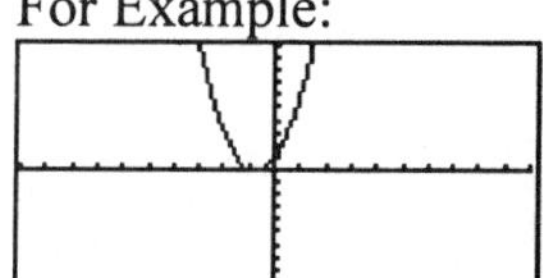

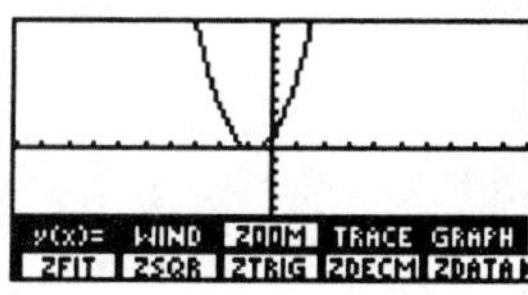

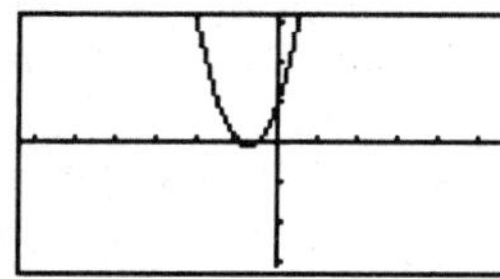

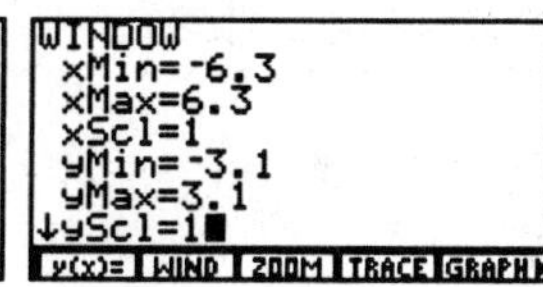